THE CORE TEST WRAPPER HANDBOOK

FRONTIERS IN ELECTRONIC TESTING

Consulting Editor
Vishwani D. Agrawal

Books in the series:

Digital Timing Measurements – From Scopes and Probes to Timing and Jitter
Maichen, W., Vol. 33
ISBN 0-387-32418-0
Fault-Tolerance Techniques for SRAM-based FPGAs
Kastensmidt, F.L., Carro, L. (et al.), Vol. 32
ISBN 0-387-31068-1
Data Mining and Diagnosing IC Fails
Huisman, L.M., Vol. 31
ISBN 0-387-24993-1
Fault Diagnosis of Analog Integrated Circuits
Kabisatpathy, P., Barua, A. (et al.), Vol. 30
ISBN 0-387-25742-X
Introduction to Advanced System-on-Chip Test Design and Optimi...
Larsson, E., Vol. 29
ISBN: 1-4020-3207-2
Embedded Processor-Based Self-Test
Gizopoulos, D. (et al.), Vol. 28
ISBN: 1-4020-2785-0
Advances in Electronic Testing
Gizopoulos, D. (et al.), Vol. 27
ISBN: 0-387-29408-2
Testing Static Random Access Memories
Hamdioui, S., Vol. 26,
ISBN: 1-4020-7752-1
Verification by Error Modeling
Radecka, K. and Zilic, Vol. 25
ISBN: 1-4020-7652-5
Elements of STIL: Principles and Applications of IEEE Std. 1450
Maston, G., Taylor, T. (et al.), Vol. 24
ISBN: 1-4020-7637-1
Fault Injection Techniques and Tools for Embedded systems Reliability …
Benso, A., Prinetto, P. (Eds.), Vol. 23
ISBN: 1-4020-7589-8
Power-Constrained Testing of VLSI Circuits
Nicolici, N., Al-Hashimi, B.M., Vol. 22B
ISBN: 1-4020-7235-X
High Performance Memory Memory Testing
Adams, R. Dean, Vol. 22A
ISBN: 1-4020-7255-4
SOC (System-on-a-Chip) Testing for Plug and Play Test Automation
Chakrabarty, K. (Ed.), Vol. 21
ISBN: 1-4020-7205-8
Test Resource Partitioning for System-on-a-Chip
Chakrabarty, K., Iyengar & Chandra(et al.), Vol. 20
ISBN: 1-4020-7119-1
A Designers' Guide to Built-in Self-Test
Stroud, C., Vol. 19
ISBN: 1-4020-7050-0

THE CORE TEST WRAPPER HANDBOOK
Rationale and Application of IEEE Std. 1500™

by

Francisco da Silva

Teresa McLaurin

Tom Waayers

Springer

Francisco da Silva
Teresa McLaurin
Tom Waayers

The Core Test Wrapper Handbook
Rationale and Application of IEEE Std. 1500™

ISBN-10: 1-4899-8769-X
ISBN-13: 978-1-4899-8769-3
ISBN-10: 0-387-34609-0 (e-book)
ISBN-13: 978-0-387-34609-0 (e-book)

Printed on acid-free paper.

Softcover re-print of the Hardcover 1st edition 2006

9 8 7 6 5 4 3 2 1

springer.com

This book is dedicated to

Alyssa, David, Diek, Earvin, Eus, Flo, Gé,

Ian, Jason, Lidwine, Mark, and Patricia.

Contents

List of Figures xv
List of Tables xxi
Foreword xxiii
Preface xxvii
Acknowledgements xxix
Introduction 1

1. What is the IEEE 1500 Standard? 3
1.1 The IEEE 1500 Wrapper Architecture 6
1.1.1 The Wrapper Interface Port 7
1.1.2 The 1500 Wrapper Instruction Register 8
1.1.3 The Wrapper Boundary Register 8
1.1.4 The Wrapper Bypass Register 9
1.2 Compliance to IEEE 1500 9
1.2.1 IEEE 1500 Unwrapped Compliance 9
1.2.2 IEEE 1500 Wrapped Compliance 10
1.3 The Core Test Language (CTL) 11
1.3.1 What is CTL? 11

2. Why use the IEEE 1500 Standard? 13
2.1 Introduction 13
2.1.1 Test Reuse and Partitioning 13
2.1.2 Partition-Based Testing, Debug and Diagnosis 17
2.2 Why Was a Standard Needed? 18

3. Illustration Example 21
3.1 Introduction 21
3.2 Unwrapped Core 21
3.3 Writing CTL For the EX Core 25
3.3.1 CTL Code Convention 25
3.3.2 CTL Code For the Unwrapped EX Core 25

4. Design of the IEEE 1500 Interface Port 29

4.1 Creation of the Wrapper Interface Port 29
4.1.1 Creation of the WSP 29
4.1.1.1 Accessing the 1500 Wrapper Through the WSP 30
4.1.1.2 Understanding WRCK... 31
4.1.1.3 Understanding WRSTN... 31
4.1.1.4 Understanding SelectWIR...................................... 32
4.1.1.5 Understanding ShiftWR 33
4.1.1.6 Understanding CaptureWR 33
4.1.1.7 Understanding UpdateWR..................................... 34
4.1.1.8 Understanding TransferDR 35
4.1.1.9 Understanding AUXCK Clocks 35
4.1.1.10 Understanding the WSI Terminal.......................... 36
4.1.1.11 Understanding the WSO Terminal 37
4.1.2 Understanding the WPP 37

4.2 CTL Template for the Wrapper Interface Port 38
4.2.1 Structure of the CTL Internal block 40
4.2.1.1 Specifying Electrical Characteristics Using the ElectricalProperty Keyword 40
4.2.1.2 Identifying 1500 Terminals Using the Wrapper Keyword 41
4.2.1.3 Specifying Test Pin Type Using the DataType Keyword 42
4.2.1.4 Specifying Input Pin Timing Using the DriveRequirements Keyword... 45
4.2.1.5 Specifying Output Pin Timing Using the StrobeRequirements Keyword 47

4.3 Wrapper Interface Port Creation for the EX Core 49
4.3.1 Declaring WRCK for the EX wrapper in serial mode 53
4.3.2 Declaring WRSTN for the EX wrapper 54
4.3.3 Declaring SelectWIR for the EX Wrapper 55
4.3.4 Declaring ShiftWR for the EX Wrapper 55
4.3.5 Declaring CaptureWR for the EX wrapper 56
4.3.6 Declaring UpdateWR for the EX Wrapper 57
4.3.7 Declaring WSI for the EX Wrapper 58
4.3.8 Declaring WSO for the EX Wrapper 58
4.3.9 Declaring WPC for the EX Wrapper 59
4.3.9.1 Declaring WRCK for the EX Wrapper in parallel mode 59
4.3.9.2 Declaring a Functional Clock for the EX Wrapper 60

4.3.9.3 Declaring WPC for the EX wrapper scan mode..... 61
4.3.9.4 Declaring WPC for the EX wrapper MBIST mode 62
4.3.10 Declaring WPI for the EX Wrapper 63
4.3.11 Declaring WPO for the EX Wrapper 65
4.3.11.1 Declaring WPO for the EX wrapper scan test mode.. 65
4.3.11.2 Declaring WPO for the EX wrapper MBIST mode 66
4.4 Combined CTL Code for EX Wrapper Terminals 68
4.5 Plug-and-play of the 1500 Architecture 75

5. **Instruction Types** 77

5.1 Overview of the IEEE 1500 Standard Instructions 77
5.1.1 Mandatory Instructions 79
5.1.1.1 WS_BYPASS 80
5.1.1.2 WS_EXTEST 80
5.1.1.3 Wx_INTEST 82
5.1.2 Optional Instructions 83
5.1.2.1 WP_EXTEST 84
5.1.2.2 Wx_EXTEST 85
5.1.2.3 WS_SAFE 85
5.1.2.4 WS_PRELOAD 86
5.1.2.5 WP_PRELOAD 87
5.1.2.6 WS_CLAMP 87
5.1.2.7 WS_INTEST_SCAN 88
5.1.2.8 WS_INTEST_RING 88
5.1.3 User-Defined Instructions 88
5.2 EX Wrapped Core Instructions 89
5.3 CTL Template for 1500 Instructions 93
5.3.1 Describing Patterns in CTL 93
5.3.1.1 Describing Pattern Data in CTL 93
5.3.2 Accessing Information Outside the CTLMode Block Using the DomainReferences Keyword 98
5.3.3 Describing Test Modes Using the TestMode Keyword 99
5.3.4 Specifying Pattern Information for Test Modes Using the PatternInformation Keyword 100
5.3.4.1 The Pattern and PatternBurst Constructs.............. 100
5.3.4.2 The PatternExec Construct 102
5.3.4.3 The Procedure and Macro Constructs 103
5.3.5 Identifying IEEE 1500 Instructions Using the TestModeForWrapper Keyword 104

5.4 Combined 1500 Instructions CTL Code Template 104
5.5 Describing CTL Code for EX Wrapper Instructions.................. 105
5.5.1 Describing WS_BYPASS in CTL for the EX Wrapper .. 106
5.5.1.1 Loading the WS_BYPASS Instruction 106
5.5.1.2 Understanding WS_BYPASS Setup Protocol and Data.. 108
5.5.1.3 Describing the EX Wrapper WS_BYPASS Setup Pattern in CTL 109
5.5.1.4 Defining the CTLMode Block for the WS_BYPASS Mode.. 110
5.5.2 Describing WP_BYPASS in CTL for the EX Core 111
5.5.2.1 Defining the CTLMode block for the WP_BYPASS Mode.. 112
5.5.3 Describing WS_EXTEST in CTL for the EX Core 113
5.5.3.1 Defining the CTLMode Block for the WS_EXTEST Mode.. 114
5.5.4 Describing WP_INTEST in CTL for the EX core 115
5.5.4.1 Defining the CTLMode Block for the WP_INTEST Mode.. 116
5.5.5 Describing WS_INTEST in CTL for the EX Core 117
5.5.5.1 Defining the CTLMode Block for the WS_INTEST Mode.. 118
5.5.6 Describing WP_INTEST_MBIST in CTL for the EX core .. 119
5.5.6.1 Defining the CTLMode Block for the WP_INTEST_MBIST Mode............................... 120
5.5.7 Describing WP_EXTEST in CTL for the EX core 121
5.5.7.1 Defining the CTLMode Block for the WP_EXTEST Mode.. 122
5.5.8 Describing WS_SAFE_SINGLE in CTL for the EX Core ... 123
5.5.8.1 Defining the CTLMode Block for the WS_SAFE_SINGLE mode 124
5.5.9 Describing WP_EXTEST_SEQ in CTL for the EX Core 125
5.5.9.1 Defining the CTLMode Block for the WP_EXTEST_SEQ mode................................... 126
5.5.10 Describing WP_INTEST_SEQ in CTL for the EX core . 127
5.5.10.1 Defining the CTLMode Block for the WP_INTEST_SEQ mode.................................... 128

6. Design of the WBR 131

6.1 What is the WBR? 132

6.2 Provision of WBR cells 133

6.3 Creation of 1500 compliant WBR cells 134

6.3.1 WBR Cell operation 135

6.3.1.1 Providing support for the shift operation 136

6.3.1.2 Providing support for the capture operation 137

6.3.1.3 Providing support for the update operation 138

6.3.1.4 Providing support for the transfer operation 139

6.3.2 Dedicated WBR Cell implementation 140

6.3.3 Shared WBR Cell implementation 143

6.3.4 Harness Cells and Reduced Functionality Cells 145

6.4 Describing wrapper cells in 1500 146

6.4.1 Describing WBR cells 147

6.4.1.1 Understanding bubble diagrams 147

6.4.1.2 WBR cell naming convention 149

6.4.1.3 Harnessing logic naming convention 150

6.5 WBR Operation Events 151

6.5.1 WBR Operation Modes 152

6.5.2 Parallel configuration of the WBR 154

6.6 Creation of the WBR for the EX example 156

6.6.1 EX core test requirements 156

6.6.2 EX core WBR cell design 157

6.6.2.1 Dedicated WBR cell assignment 157

6.6.3 Shared WBR cell design 160

6.6.4 Special WBR cell design 162

6.6.4.1 WBR cell for Bus Control 162

6.6.4.2 WBR cell for Reset 164

6.6.5 WBR design 167

6.6.5.1 Parallel configuration of the WBR 168

6.7 Describing the WBR in CTL 170

6.7.1 Describing WBR cells in CTL at a Wrapper Level 171

6.7.2 Describing WBR Cells in CTL at a Core Level 172

6.7.3 Describing Scan Chains in CTL 173

6.8 Describing the WBR chain of the EX wrapper in CTL 174

6.8.1 EX wrapper serial mode WBR chain 174

6.8.2 EX wrapper parallel mode WBR chains 175

6.9 Describing WBR Cells of the EX Wrapper in CTL 176

7. **Design of the WBY** 181

7.1 Introduction 181

7.2 WBY Construction 181

7.3 WBY Utilization 184

7.4 WBY for the EX Core 185

7.5 Describing the WBY in CTL for the EX Core 185

8. **Design of the WIR** 187

8.1 WIR architecture 187

8.1.1 WIR Operation 188

8.2 Understanding the WIR Rules 190

8.2.1 WIR Configuration and DR selection 190

8.2.2 Wrapper Disabled and Enabled States 191

8.2.3 WIR Design 191

8.3 Creation of the WIR for the EX example 196

8.3.1 Design of WIR building blocks 197

8.3.2 EX core WIR design 198

8.3.3 Design of WIR circuitry for 1500 register control 203

8.3.3.1 WIR control circuitry 203

8.3.3.2 WBY control circuitry 203

8.3.3.3 WBR control circuitry 204

8.3.3.4 Core control circuitry 209

8.3.4 Design of WIR circuitry for register configuration 211

8.3.4.1 Parallel port register configuration 212

8.3.4.2 Serial port register configuration 214

8.3.5 WIR circuitry for clock selection 215

8.4 Describing the WIR in CTL for the EX Core 216

9. **Hierarchical Cores** 217

9.1 Dealing with Hierarchical Cores 217

9.2 WIR in a Hierarchical Core 218

9.3 WBY in a Hierarchical Core 218

9.4 WBR in a Hierarchical Core 220

10. **Finalizing the Wrapper Solution for the EX Core** . 221

10.1 Expressing Compliance to the 1500 Standard 221

10.1.1 Expressing 1500 compliance for the EX wrapper 222

10.2 The Wrapped Core 222
10.2.1 Wrapping the EX Core 222
10.3 Defining the wrapper reset operation in CTL 223
10.4 Pattern examples........ 225
10.4.1 WS_INTEST pattern example 225
10.4.2 WP_INTEST pattern example 227
10.5 Combined CTL Code for the EX Wrapper 229

11. SOC Integration of 1500 Compliant Cores 249

11.1 SOC integration........ 249
11.1.1 1500 compliant wrapped core types. 249
11.1.2 Core connection types 251
11.1.3 Connecting 1500 compliant cores 253
11.1.4 Interfacing to SOC pins 257
11.2 TAP controller interface........ 261
11.2.1 Interfacing TAP FSM to WSC 262
11.2.2 Mapping TAP states to WIR events 264
11.2.3 Mapping TAP states to WBR cell events 266
11.2.3.1 Mapping TAP states to the WBR cell Transfer event 266
11.3 Scheduling considerations........ 268

Conclusion 273

Index 275

List of Figures

Figure 1	Architecture of a 1500-wrapped core	7
Figure 2	Isolated Core with Pattern	14
Figure 3	Core Embedded in an SOC	15
Figure 4	MUXES Block	16
Figure 5	Core Pattern at SOC Level	17
Figure 6	Simultaneous Access of Two Wrappers	19
Figure 7	Unwrapped Core Example: EX	22
Figure 8	Example of Registered and Unregistered Ports	22
Figure 9	High-level CTL test mode structure	27
Figure 10	WRCK timing waveform	31
Figure 11	WRSTN timing waveform - asynchronous implementation	32
Figure 12	Timing waveform of the 1500 SelectWIR function	33
Figure 13	Timing waveform of the 1500 ShiftWR function	33
Figure 14	Timing waveform of the 1500 CaptureWR function	34
Figure 15	Timing waveform of the 1500 UpdateWR function	34
Figure 16	Timing waveform of the 1500 TransferDR function	35
Figure 17	Timing waveform of the 1500 AUXCK terminal	36
Figure 18	Timing waveform of the 1500 WSI terminal	36
Figure 19	Timing waveform of the 1500 WSO terminal	37
Figure 20	Wrapper design level for the EX core	51
Figure 21	Timing waveform of the EX wrapper WPSE terminal	61
Figure 22	Timing waveform of the EX wrapper MBIST WPC terminals	62
Figure 23	Timing waveform of the WPSI terminal	64

Figure 24 Timing waveform of the WPSO terminal ... 65
Figure 25 Timing of the EX wrapper MBIST WPO terminals ... 66
Figure 26 WS_EXTEST Example ... 81
Figure 27 WBR cell examples using WSC or WPP ... 85
Figure 28 Example of a WBR cell with a safe gate ... 86
Figure 29 WC_SD1_COI_UD WBR Cell ... 87
Figure 30 WBR chain(s) concatenated with internal chain(s) ... 88
Figure 31 Utilizing WP_BYPASS ... 91
Figure 32 EX wrapper WS_BYPASS instruction load timing ... 108
Figure 33 EX wrapper WP_BYPASS instruction load timing ... 112
Figure 34 EX wrapper WS_EXTEST instruction load timing ... 114
Figure 35 EX wrapper WP_INTEST instruction load timing ... 116
Figure 36 EX wrapper WS_INTEST instruction load timing ... 118
Figure 37 EX wrapper WP_INTEST_MBIST instruction load timing .. 120
Figure 38 EX wrapper WP_EXTEST instruction load timing ... 122
Figure 39 EX wrapper WS_SAFE_SINGLE instruction load timing 124
Figure 40 EX wrapper WP_EXTEST_SEQ instruction load timing 126
Figure 41 EX wrapper WP_INTEST_SEQ instruction load timing 128
Figure 42 Illustration of external logic test via 1500 wrappers ... 131
Figure 43 Example serial and parallel configuration of a WBR ... 133
Figure 44 WBR cell data terminals ... 135
Figure 45 WBR cell with multiple storage elements on shift path ... 137
Figure 46 Example WBR cell with capture from CFI ... 137
Figure 47 Example WBR cell with capture from CFO ... 138
Figure 48 Example WBR cell with capture in last storage element ... 138
Figure 49 Example WBR cell with an update stage ... 139
Figure 50 Example WBR cell supporting transfer ... 140
Figure 51 Registered and unregistered functional core terminals ... 141
Figure 52 Dedicated wrapper cell example ... 141
Figure 53 Example of dedicated wrapper cell operations ... 142
Figure 54 Example dedicated wrapper cell usage on registered ports ... 143
Figure 55 Example of shared wrapper cell ... 144
Figure 56 Example shared wrapper cell operations ... 144

Figure 57 Example shared wrapper cell usage 145
Figure 58 Pass-Through Harness Cells 145
Figure 59 WC_SD1_CN_G cell 146
Figure 60 Example implementation WBR cell: WC_SD1_CII_UD 147
Figure 61 WBR cell operation bubble diagram representations 148
Figure 62 Bubble diagram for WBR cell: WC_SD1_CII_UD 149
Figure 63 WBR cells in Inward Facing mode 152
Figure 64 WBR cells in Outward Facing mode 153
Figure 65 Example parallel configuration of core chains and WBR 155
Figure 66 Implementation WC_SD1_COI 157
Figure 67 Bubble diagram WC_SD1_COI 159
Figure 68 Implementation WC_SF1_CII 160
Figure 69 Example implementation WC_SF1_CII 161
Figure 70 Bubble diagram WC_SF1_CII 161
Figure 71 Implementation WC_SD1_COI_G for Bus Control 162
Figure 72 Bubble diagram WC_SD1_COI_G 164
Figure 73 External Reset Logic Example 164
Figure 74 Implementation WC_SD1_CII_O 165
Figure 75 Bubble diagram WC_SD1_CII_O 167
Figure 76 WBR for the EX core 169
Figure 77 Parallel configuration of the WBR for the EX core 170
Figure 78 WBY Example 182
Figure 79 WBY mechanism 182
Figure 80 WBY mechanism with gated WSI 183
Figure 81 WS_SAFE example 183
Figure 82 WBY Example 184
Figure 83 EX Core WBY register 185
Figure 84 Example WIR 188
Figure 85 Protocol for resetting the WIR and loading an instruction 189
Figure 86 Example WIR design with asynchronous reset 192
Figure 87 Example WIR Design with Synchronous Reset 194
Figure 88 Example WIR Building Block 195
Figure 89 Building block WIR shift register 197

Figure 90 Building block WIR update register 198
Figure 91 Implementation of the WIR for the EX core 199
Figure 92 Implementation of the WIR circuitry for the EX core 202
Figure 93 WIR control circuitry 203
Figure 94 WBY control circuitry 203
Figure 95 WBR control circuitry: hold_inputs 204
Figure 96 WBR control circuitry: hold_outputs 206
Figure 97 WBR control circuitry: wse_inputs 207
Figure 98 WBR control circuitry: wse_outputs 208
Figure 99 WBR control circuitry: bus_disable 209
Figure 100 WBR control circuitry: se 209
Figure 101 Core static control circuitry 210
Figure 102 Parallel Port register configuration 212
Figure 103 EX wrapper WBR bypass segment 213
Figure 104 Serial Port register configuration 214
Figure 105 Logical representation of the EX core clock source selection 215
Figure 106 Hierarchical Core 217
Figure 107 Hierarchical WIR Path 218
Figure 108 Embedded 1500 Core with non-hierarchical WBY path 219
Figure 109 Embedded 1500 Core with Hierarchical WBY path 219
Figure 110 1500 Component Connections 223
Figure 111 WS_INTEST patterns 226
Figure 112 WP_INTEST patterns 228
Figure 113 SOC with three types of 1500 compliant wrapped cores 250
Figure 114 Example 1500 compliant wrapper interfaces 250
Figure 115 Interface connection types 252
Figure 116 WSP distribution and daisy-chaining 254
Figure 117 WPP distribution 255
Figure 118 WPP daisy-chaining 256
Figure 119 WPP Multiplexing 256
Figure 120 WPP width adaptation 257
Figure 121 Reusing functional pins for WSP access 258
Figure 122 Single TAP controller WSP access 259

Figure 123 Reusing functional pins for WPP access 260
Figure 124 Using the WIR for SOC test control 260
Figure 125 Multiple TAP controller architecture 261
Figure 126 IEEE Std. 1149.1 TAP FSM states 262
Figure 127 Mapping TAP states to WSC control 263
Figure 128 WIR events when controlled as JTAG data register 265
Figure 129 WIR events when controlled as JTAG Instruction Register .. 266
Figure 130 WBR cell events when controlled as JTAG data register 266
Figure 131 Example WBR cell Transfer event when controlled as JTAG data register 268
Figure 132 Example test schedule with WIR initialization and WPP protocols 269
Figure 133 Example test schedule with parallel WPP protocols 270
Figure 134 Example test schedule with WPP protocols and WIR initialization overlapping 271
Figure 135 Example test schedule with sequence of WPP and WSC protocols 271
Figure 136 Example of test schedule with WPP and WSC protocols in parallel 272

List of Tables

Table 1	Core Port to SOC Connection	16
Table 2	UnWrapped Core Pins	23
Table 3	Core Modes	24
Table 4	DataType ActiveState arguments and corresponding logic value mapping	44
Table 5	IEEE 1500 Instruction List	79
Table 6	IEEE 1500 Optional Instructions	84
Table 7	EX Core instruction definitions	92
Table 8	EX wrapper instruction opcodes	106
Table 9	Example of dedicated wrapper cell operation control	142
Table 10	Example of shared wrapper cell operation control	144
Table 11	Control values for wrapper configuration in parallel mode	156
Table 12	Control values for WC_SD1_COI	158
Table 13	Control values for WC_SF1_CII	160
Table 14	Control values for WC_SD1_COI_G	163
Table 15	Control values for WC_SD1_CII_O	166
Table 16	Control values for the WBR chain	167
Table 17	WBR cell name and type mapping to EX wrapper terminals	176
Table 18	WIR Building block Control Values	196
Table 19	WIR Instruction opcodes and output values for EX wrapper	200
Table 20	WIR control values for WBR control: hold_inputs	205
Table 21	WIR control values for WBR control: hold_outputs	206
Table 22	WIR control values for WBR control: wse_inputs	207

Table 23 WIR control values for WBR control: wse_outputs 208
Table 24 WIR control values for core control: se 210
Table 25 WIR control values for Core static control signals 211
Table 26 WIR control values for Register configuration 214
Table 27 WIR control values for Functional clock source selection 216

Foreword

In the early to mid-1990's while working at what was then Motorola Semiconductor, business changes forced my multi-hundred dollar microprocessor to become a tens-of-dollars embedded core. I ran into first hand the problem of trying to deliver what used to be a whole chip with something on the order of over 400 interconnect signals to a design team that was going to stuff it into a package with less than 220 signal pins and surround it with other logic. I also ran into the problem of delivering microprocessor specification verification – a microprocessor is not just about the functions and instructions included with the instruction set, but also the MIPs rating at some given frequency. I faced two dilemmas: one, I could not deliver functional vectors without significant development of off-core logic to deal with the reduced chip I/O map (and everybody's I/O map was going to be a little different); and two, the JTAG (1149.1) boundary scan ring that was around my core when it was a chip was going to be woefully inadequate since it did not support at-speed signal application and capture and independent use separate from my core. I considered the problem at length and came up with my own solution that was predominantly a separate non-JTAG scan test wrapper that supported at-speed application of launch-capture cycles using the system clock.

But my problems weren't over at that point either. Being a core-provider was one thing, however, I was also in the business of being a core-integrator. This meant receiving cores from others, and I noticed that no two cores had the same vector, vector delivery and test solutions. Some cores relied on built-in self-test; some cores relied on functional vectors applied through multiplexors from the chips I/O map; some cores relied on functional vectors turned into serial vectors and shifted in through an 1149.1 compliant boundary-scan ring; and some cores had no solution delivered at all, just the functional vectors that were used for verification, or the scan vectors that were used when the core was a standalone chip. I thought about this problem as well and the best solution that I could apply was to create a specification and to accept from my core-providers only those cores that met my exact requirements. Being that the IP core-business was in it's infancy – this would have had the effect of limiting my ability to get cores. I did talk to my core providers and they said to me what I had said to others previously "that everybody wants something different." Since IP cores were to be bought, sold, and traded among different companies, it was quickly apparent to those in the business that everyone having their own specific or even proprietary solution for delivering cores or requiring their own solution for accepting cores, was going to result in chaos (my joke at the time was that it would require two lawyers for every engineer to get this business to work – I didn't realize how close I was to the truth).

When the P1500 movement started, there were a great many interested people. One of the early meetings at ITC in Washington, D.C., filled a room with a full complement of very opinionated people. I knew then that this was going to lead to a lengthy development cycle, but that every opinion would get aired and the result would be a full-featured solution. As it was, a great many people passed through the working group (many of whom I have worked closely with over the years) and it took seven years to come to the point where the standard was voted on and ratified. As expected, the final result is a full-featured well-constructed and usable solution that is already available from library and EDA vendors.

One of the lessons I learned from my close work with JTAG (IEEE Standard 1149.1) in the past, was that the way standards are written, based largely on rules and permissions and measuring compliance, is not the way people that want to learn the standard wish to get the information presented to them. I actually worked on some 1149.1 parts in the late 1980's before the JTAG standard was fully ratified and I understood all of the principle features. However, when I finally read the completed standard, I was amazed at how convoluted the information was presented. In that same vein, a handbook or layman's explanation of the technology, application and use cases of a standard is almost required for widespread adoption. This book is that handbook for the IEEE 1500 standard. The material is presented in a manner much more suited for learning (rather than trying to integrate a box of rules in your head) and from a practical point of view. The authors are some of the most knowledgeable people available since they are members of the working group that saw the ratification of their long efforts into a viable and useful standard.

The book begins by covering the concept of just what the 1500 IEEE Standard represents, and then moves from there to answering the question, "Why use the IEEE 1500 Standard?". The use of the 1500 Standard is then described through an illustrated example; and the steps of designing the interface port, the boundary register, the bypass register, the instruction register, and the instructions all receive their own individual chapters. A good amount of material to familiarize the reader with the description language, CTL (Core Test Language) and compliance is provided – and then the complete wrapping of a core is described. The final chapter discusses the integration of 1500 cores into a System-On-Chip (SOC) design.

It doesn't seem that core-based design will slow down anytime in the near future, design windows are continually shrinking, and more complex integrations make portability, reuse, and schedule-efficiency the key drivers to high-quality and time-to-market. 1500 will be part of this landscape, which hopefully will make this book part of every DFT designer's library.

ALFRED L. CROUCH
Chief Scientist
Inovys Corp.

Preface

Standards tend to get molded by a very normative structure with assumed hierarchical relationships between rules, recommendations, and permissions. While this structure adequately suits standardization needs, it often poses interpretation challenges and fails to easily convey the intent behind some of these rules, recommendations, and permissions. In addition, hardware standards tend to be generic in nature and laconic with respect to hardware implementation discussion because this is, typically, outside their scope. As a result, a non-formal post-standardization effort is often useful to the adoption of new standards, and is the place where a less-constrained presentation of the standards' prescriptions can take place and where implementation examples can be offered freely. 1500 is no exception to this and in fact one could argue that it faces the additional adoption hurdle of combining its hardware architecture with another IEEE language standard (1450.6 CTL) which is also in the early stages of its adoption ramp. This book is meant to help address these challenges.

The following chapters will take the reader through an in-depth understanding of the rationale behind the 1500 architecture, and the resulting knowledge will be applied to an example core design chosen to illustrate application of the 1500 standard. Specific 1500 wrapper hardware logic and corresponding CTL code will be progressively added to this example design throughout the book, complementing the more theoretical discussion of the 1500 architecture. Whenever appropriate, each chapter will begin with a general discussion of the relevant characteristics of the standard and end with presentation of the corresponding 1500 wrapper component for the illustration example. A deliberate emphasis on CTL will be noticeable throughout the book as this language is indispensable to the process leading to 1500 compliance. This emphasis might appear too verbose at first but it should be noted that, ultimately, the CTL code associated with 1500 cores is meant to be created and consumed by automation tools. While this book is not intended to provide CTL expertise, it does offer the basic understanding of CTL that required to enable compliance to IEEE 1500.

Beyond compliance with 1500, one has to also wonder about integration of 1500 compliant cores at the SOC level. This is outside the scope of 1500 but remains an indispensable step in the test of 1500 cores. The last chapter of this book addresses this issue by discussing SOC integration of these cores, and also the creation of an interface between 1500 and 1149.1 at the SOC level. This final chapter also includes scheduling considerations at the SOC level.

Acknowledgements

The 1500 standardization activity has required a lot of personal contribution and tenacity from many people over the years. We would like to recognize the weekly commitment that our colleagues from the 1500 scalable architecture task force, the 1500 compliance definition task force, and the 1500 editorial task force have dedicated to the completion of the 1500 standard.

In addition, we would like to thank Ed van Veldhoven (Holland Vormgevers) for his cover design, and Ben Bennetts, Grady Giles, and Rohit Kapur for their invaluable review feedback, and Al Crouch for his feedback and insightful stage-setting words.

Francisco da Silva

Teresa McLaurin

Tom Waayers

Introduction

The IEEE 1500 working group took about seven years to complete definition and approval of IEEE Std. 1500 in 2005. While many would argue that this was a long time, it takes a close look at the resulting standardization work to fully grasp the scope of the working group's accomplishments. The 1500 hardware architecture was designed with a significant amount of flexibility that should allow the user to define and implement wrappers that fit a variety of test requirements including sequential tests. IEEE 1500 is in many ways similar to the 1149.1 architecture but does not overlap with the JTAG standard because, unlike 1149.1, 1500 is a block-level or core-level solution. Careful consideration was employed during the development of the 1500 standard to maintain compatibility with 1149.1 for the obvious reason that 1500 is not a chip-level solution and therefore needs to be combined with a chip-level architecture so as to enable testing of the 1500 wrapped core in silicon. Besides the definition of a scalable hardware architecture, the 1500 activities initiated specification of a new test language (CTL - Core Test Language) which has now acquired reaches far beyond 1500 core wrapper needs. Although CTL is now a distinct IEEE standard (1450.6), its original core test aspects created for 1500 needs remain and are still relied upon by 1500. Both the hardware architecture and the CTL are indispensable components of the IEEE 1500 standard.

Chapter 1
What is the IEEE 1500 Standard?

When the IC design community was confronted with ever-shrinking design cycles, dictated by aggressive Time-To-Market, it reacted by optimizing its design productivity. Out of the many solutions that it devised, one consisted of identifying design functions that were identical across chip projects, isolating them, implementing them only once, and then reusing them across chips. The design reuse phenomenon was then born and aggressive Time-To-Market goals were made achievable, even on complex SOC designs.

This, however, came with an underlying shift in design methodologies. Indeed, for design reuse to be efficient, the chip designer had to be relieved from the design of these common-across-chip design functions (referred to as cores), and essentially become a core user while a separate entity became the core provider.

It did not take long before a follow-up design cycle optimization necessity surfaced: If a design is being reused then the corresponding DFT and pattern generation efforts should also be leveraged by the core user. This seems natural, but does come with specific test related challenges: How will the core provider's DFT strategy fit into the core user's SoC environment, when core providers and core users are different entities and when design terminals which were primary-inputs at the core level become embedded at the SOC level? The technical challenge associated with this issue was one of controllability and observability of these cores at the SOC level; and it became apparent that DFT logic had to be inserted to enable efficient test reuse on these cores at the SOC level. Beyond this technical challenge, there was a communication challenge as well; one that relates to establishing accurate and complete transfer of test related information between core providers and core users, who were often different entities, so that their DFT strategies can be aligned. So how were these problems solved?

Driven, yet again, by Time-To-Market, many companies adopted ad-hoc solutions to these problems. The consequence was that core users were confronted with a variety of custom solutions in their interactions with core providers. Although, these solutions were not identical across companies, they all attempted to address the same technical challenge which is that of providing controllability and observability to embedded cores at the SOC level. The solution consisted of:

1. making use of SOC level routes – called Test Access Mechanisms (TAMs) – to transport test data from the SOC pins to the embedded core boundary, and
2. de-serializing the test data coming from often reduced-bandwidth TAMs so that this data can be applied to cores that have a higher bandwidth. Also, high-bandwidth core outputs had to be serialized to match the reduced bandwidth of the SOC TAMs. The design structure allowing for this bandwidth adaptation at the core interface is part of DFT logic inserted around the core and called a wrapper.

The problem that some of these custom solutions faced was that they were inefficient. Some solutions unnecessarily increased TAM width, hence creating routing and other SOC level resource issues. Others were inefficient because they unnecessarily increased the core's test time by supporting only a serial type of test where core test time increased in proportion with the length of their wrapper chain. These performance related problems had to be solved in order to make core-based testing efficient.

Besides these implementation specific shortcomings, the wrapper approach remained a viable solution to the issue of providing controllability and observability to embedded cores. Many companies were able to design efficient implementations of the wrapper. However, these companies were left with the burden of establishing their own communication mechanism to ensure accurate and complete alignment of DFT strategies between core providers and core users. These issues made apparent the need for standardization, not only on the creation of an efficient wrapper structure, but also on the communication of the characteristics of this wrapper from the core provider to the core user. For instance, how would the core user be informed of the fact that a particular core was wrapped with wrapper cells implemented with multiple storage elements so as to allow a sequential type of test on the wrapped core, independently of the origin of the core? How would the core user learn that a particular wrapper level terminal is a wrapper scan input pin and therefore needs to be connected in a certain way at the SOC level? How would the core user be able to handle the format in which core level patterns are delivered by the core provider – again independent of the origin of the core?. These are just some of the questions for which the IEEE 1500 working group was created to address, with an overall goal of enabling efficient core-based testing at the SOC level.

One of the guiding principles of the 1500 standardization activity was the fact that different companies and designs need different DFT strategies -

requiring stuck-at, transition-delay, path-delay, functional, isolation when testing a particular core, quiescence or isolation when testing others, and test coordination between cores. Therefore, standardization was to be avoided with regard to DFT strategies, but had to be focused on provision of a flexible DFT framework. This framework had to support the diverse DFT strategies that can be implemented by the core provider and that have to be accurately and completely expressed to the core user. Another characteristic of a standardized core test solution would be to enable efficient reuse, through SOC TAMs, of test data created at the core level. This would allow core providers to make DFT patterns available with their cores independent of their test strategies.

The 1500 standard addressed these issues by defining a flexible and scalable DFT architecture that supports a variety of test strategies and uses IEEE Std. 1450.6 – Standard for Standard Test Interface Language (STIL) for Digital Test Vector Data - Core Test Language (CTL) – to solve communication needs between core providers and core users. The 1500 architecture consists of:

a. standard core wrapper components that meet both combinational and sequential core test strategies, and

b. CTL requirements that allow for the characteristics of the wrapper to be communicated to the core user and ultimately allow for this information to be used to efficiently integrate the core and map core level test data to the SOC level (independent of the origin of the core).

Beyond the core provider and core user business model, the 1500 architecture offers another advantage. It allows for SOC level DFT strategies to be sub-divided and tackled from a core (or sub-design) perspective. This enables test generation at the block level and the porting (through the 1500 architecture) of the resulting core-level test patterns to the SOC level. The result is that the 1500 architecture provides an opportunity to transcend hardware and software capacity issues, as well as test pattern generation runtime issues synonymous with Moore's law.

The following subsections will provide an overview of the 1500 wrapper architecture and CTL requirements that relate to compliance with the 1500 standard.

1.1 The IEEE 1500 Wrapper Architecture

A high level understanding of the 1500 wrapper architecture requires understanding of the functions of a wrapper. A wrapper is DFT logic inserted on a core to provide an interface to the SOC level TAM after the core is integrated into an SOC. The wrapper can serve as a bandwidth adaptation mechanism between a TAM and a core given that TAMs and cores often have different bandwidth characteristics. A wrapper is also an isolation mechanism, not only for the embedded core, but also for logic surrounding the core. Lastly, the variety in core test needs requires support of the application of different types of test data through the wrapper, as well as a control mechanism that directs the type and sequence of tests applied to the embedded core. This translates into a variety of configurations or test modes in which the wrapped core must be placed in order for the test corresponding to these modes to be applied. The various wrapper functions are supported through 1500 wrapper hardware components that enable embedded core testing. As shown in Figure 1 (page 7), these wrapper components are:

- The wrapper interface port which provides an interface between the wrapper and SOC level TAMs. This interface has the flexibility of supporting various test bandwidth requirements.
- The Wrapper Instruction Register (WIR), which configures the wrapper into test modes and initiates all test activities inside the wrapped core.
- The Wrapper Boundary Register (WBR), which serves as an isolation mechanism with a data serialization and de-serialization scan chain. Various types of test data can be applied to the embedded core terminals through the WBR. Just like the wrapper interface port, the WBR can be designed to meet various test bandwidth requirements resulting in test time savings.
- The Wrapper BYpass register (WBY), which provides a short path through a wrapped core. The WBY is utilized when data meant for logic outside the core needs to traverse the core in as few clock cycles as possible so as to not incur unnecessary test time increase.

The various 1500 wrapper registers fall into three categories:

1. The Wrapper Instruction Register (WIR)
2. Wrapper Data Registers (WDRs) comprising the WBR and WBY as well as any other user-defined wrapper-level register – note that the WIR is not a WDR

3. Core Data Registers (CDRs) referring to registers inside the core wrapped by the 1500 architecture.

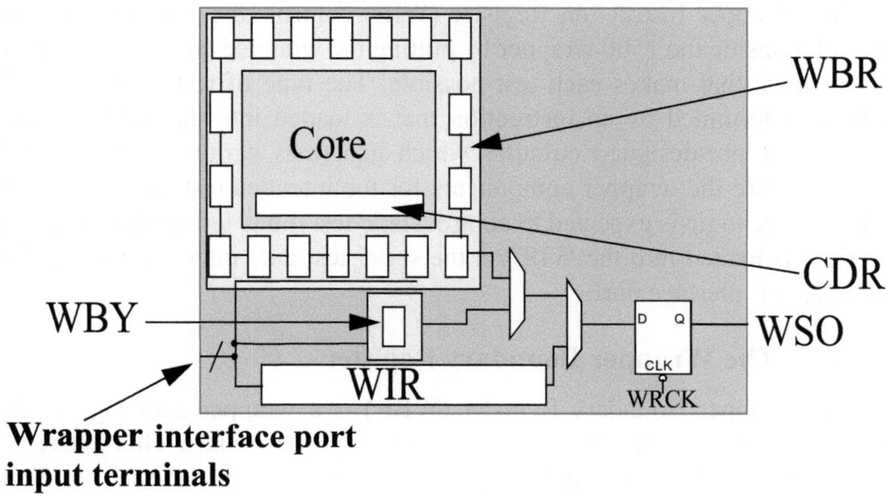

Figure 1 Architecture of a 1500-wrapped core

1.1.1 The Wrapper Interface Port

The wrapper interface port is a collection of control and data terminals that allow operation of the wrapper through a TAM, from chip-level pins or from a chip-level control entity. The wrapper interface port can provide serial and parallel access to the wrapped core, so as to meet various core test bandwidth requirements. However, 1500 focuses more on the serial interface and leaves the definition of the parallel interface to the user of the standard. The reason for this choice is that the 1500 architecture is meant to be a plug-and-play architecture, meaning that multiple 1500 compliant cores can operate together when connected through the 1500 data interface. It is difficult to achieve the plug-and-play objective with a parallel data interface. The challenge in doing so would be to define how much parallelism every core needs to have so as to create a data interface match from the output of one wrapper to the inputs of the next one. Data interface bandwidth is clearly a core-specific characteristic and therefore poses challenges for standardization efforts. Another way to look at this is that the minimum parallelism that guarantees plug-and-play is a parallelism limited to one signal. This defines a serial data interface. As a result, because every 1500 compliant core must guarantee plug-and-play, 1500 mandates a serial test data interface on every compliant wrapper. An optional parallel interface is allowed but if provided, this parallel interface must exist in addition to the serial interface.

1.1.2 The 1500 Wrapper Instruction Register

The Wrapper Instruction Register (WIR) controls the start of every test operation inside the 1500 wrapper by putting the wrapped core into the appropriate mode that makes each test possible. The type of test enabled by the WIR is determined by an instruction that is loaded into the WIR and then decoded by pre-designed circuitry which generates various control signals that configure the wrapper components for the intended test. In addition, the WIR decode logic is expected to control static test mode terminals of the core.

Data is loaded into the WIR via the serial test data interface provided on the wrapper interface port.

1.1.3 The Wrapper Boundary Register

The Wrapper Boundary Register (WBR) is a wrapper data register that provides data access to the core input and output terminals. This WDR is a collection of 1500 compliant wrapper cells provisioned for each unidirectional core terminal. Bi-directional core terminals are explicitly excluded from this support as they constitute bad practice for design reuse in general. However, 1500 wrapper cell isolation may be implemented on these terminals through wrapping of the unidirectional functions comprising the bi-directional port.

In order to support the variety of core test strategies, there is a plethora of 1500 WBR cells conceived to fulfill core and UDL test requirements. These cells are made of storage elements and receive data from the data terminal(s) of the wrapper interface port. Both a serial and a parallel configuration of the WBR are allowed in order to handle the design-specific test bandwidth requirements supported by the wrapper interface port. As a result, use of a parallel wrapper interface would typically correspond to use of a parallel configuration of the WBR. Internal testing of 1500 wrapped cores and test of the logic external to the core is enabled through an internal test configuration and an external test configuration of the WBR called Inward Facing and Outward facing modes, respectively. These modes will be explained in Section *6.5.1* (see page 152).

The 1500 architecture allows for the definition of user-defined wrapper data registers in addition to, but not in place of, the WBR.

1.1.4 The Wrapper Bypass Register

The plug-and-play characteristic of the 1500 wrapper enables connection of multiple 1500 wrapped cores at the SOC level – this will be discussed in Section *11.1.2* (see page 251). One of the advantages in serially connecting 1500 wrappers at the SOC level is the optimization of SOC test time through scheduling, as discussed later in Section *11.3* (see page 268). As a result of serially connecting multiple 1500 wrapped cores, the test data meant for a given core may need to first traverse another core or group of cores. Without the WBY, this would imply that data would be shifted through every WBR chain that exists in the path from the SOC inputs to the destination wrapper. In this scenario. the WBR would be the only mandatory data register present in the wrapper. This can create a bottleneck and an adverse test time that is proportional to the number of WBR chains that must be traversed. The role of the Wrapper BYpass register (WBY) is to provide a short route through each 1500 wrapper so as to avoid unnecessary traversing of lengthy data registers.

1.2 Compliance to IEEE 1500

A principal characteristic of the 1500 standard is that it is not a chip-level solution but a core-level one. This implies that an SOC integration must follow the application of the 1500 standard because, ultimately, test is a chip-level exercise. The core level aspect of 1500 also implies that successful use of the 1500 architecture relies on a successful SOC integration of 1500 compliant cores. It is therefore critical that complete and precise information is propagated to the core integrator, and this poses a challenge because core providers and core users are often different entities. To accommodate the core provider and core user business model, the 1500 working group defined a dual-level of compliance to the standard:

- the unwrapped compliance
- the wrapped compliance

This sub-division is meant to provide flexibility in the use of 1500, given that different design groups may be involved in the process of reaching 1500 compliance, each providing incremental contribution to this process.

1.2.1 IEEE 1500 Unwrapped Compliance

A practical issue in the design of core wrappers is that not all wrapper component decisions can be made by the core provider. For example, the type of wrapper isolation needed at the wrapped core level can be a function of the SOC environment that will host the wrapped core. The best stage to make a decision as to the type of wrapper utilized may be at the SOC integration stage. The unwrapped compliance level addresses this issue by allowing parts of the 1500 wrapper requirements to be met by the core provider and letting the core user complete the final compliance or wrapped-compliance requirements. A condition for this to be successful is that the core provider must precisely identify the core components that are meant to be reused for wrapper insertion by the SOC integrator. To satisfy this condition, 1500 requires that the core-provider exercise the wrapper insertion exercise and provide a CTL example of a 1500-wrapper that can be inserted around the core. The requirement to provide an example CTL is also meant to give the core provider a validation step that will ensure that all wrapper hardware elements that fall under the responsibility of the core provider are properly inserted. This step and the resulting CTL are what differentiate a 1500 non-compliant core from a 1500 unwrapped compliant core. Typical applications of the unwrapped compliance level revolve around the partial insertion of 1500 compliant wrapper components. However, the main criteria that enables unwrapped compliance is the provision of 1500 compliant CTL. A bare core void of any 1500 wrapper component can achieve the unwrapped compliance level with the condition that a verified CTL description is provided for each missing wrapper component.

In short, the unwrapped compliance level is meant to be an interim step towards the wrapped compliance level and existence of a 1500-compliant CTL provides certification for this step.

1.2.2 IEEE 1500 Wrapped Compliance

The 1500 wrapped compliance level is the ultimate compliance goal for the 1500 architecture. This level of compliance can be achieved starting from an unwrapped compliant core or from a non-compliant core; a non-compliant core being a core without a 1500-compliant CTL, independently of what 1500 wrapper components may exist or lack around this core. The wrapped compliance level comprises complete 1500 wrapper insertion and CTL creation.

Note: Delivering core level patterns along with an unwrapped compliant core that does not have a WBR would require pattern porting from the core level to the wrapper level and then to the SOC level. This mode of core and pattern delivery is best suited for cases where automation tools are available to handle the various pattern porting steps required on the bare core patterns. Unless such automation tools are available, it would be preferable to create the WBR and generate core-level patterns at a design level that includes the WBR. The resulting patterns and unwrapped compliant core would require less pattern protocol manipulation for the SOC level than the case where no WBR exists prior to core-level pattern generation. The same is true for fully wrapped cores as pattern generation at the wrapped core level alleviates the need for automation tool and pattern protocol manipulation for the SOC level. Some level of pattern porting from the wrapped core level to the SOC level may still be required depending on the SOC environment.

1.3 The Core Test Language (CTL)

CTL, defined by IEEE Std. 1450.6, was originally conceived to satisfy the need to communicate 1500 wrapper information between core providers and core users, and also to facilitate test data porting from the core level to the SOC level. However, this language expanded in scope and became a more generic mechanism for expression of test data, while maintaining its original ties with the 1500 architecture.

Due to its original core test application purpose, CTL plays a pivotal role in the 1500 architecture. Simply put, in order to achieve 1500 compliance, provision of a CTL code for the 1500 wrapper is as important as provision of the hardware logic that forms this wrapper. Ultimately, the CTL code corresponding to a 1500 wrapped core is meant to be created and used by automation tools designed to support embedded core test. However, understanding of the 1500 infrastructure would be incomplete if it was limited to the hardware requirements only.

1.3.1 What is CTL?

CTL, IEEE Std. 1450.6, is a STIL (pronounced “style”) based language which takes roots in IEEE Std. 1450.0 but uses STIL language extensions established in IEEE Std. 1450.1, to define a practical approach to the representation of test information. CTL has the ability to model test characteristics about designs in a way that allows this information to be used for test purposes, even in the absence of the original design. This means that core wrapper and core test information can be modelled by core providers in CTL and this CTL model can be sent to core users without disclosing any RTL or gate-level description of the core. This black-box approach suits the core design and reuse business model since these cores are often considered Intellectual Property (IP) that must be protected.

Another key aspect of CTL is that it is STIL-based and therefore has the ability to describe test patterns. In doing so, CTL natively separates test data from test protocol. Before going further it would be appropriate to look at the difference between test data and test protocol. Test information can typically be separated into two parts, the test data which gets applied to logic inside the core, and the protocol, or how to apply this data. The protocol can be defined as a sequence of events that enable the application of test data.

In the context of embedded core testing, once a wrapper is inserted, the only way to apply values on primary inputs of the core is by going through the wrapper first. This typically means that data needs to be shifted into the wrapper chain's scan input and make its way to the intended primary input cycle by cycle. The process of putting the wrapper cells into shift mode and applying the wrapper clock as many times as necessary constitutes a portion of test protocol. Why is this important? A key consideration in the embedded core test approach is the reuse, at the SOC level, of the test that was created at the core level. If the core-level test existed in a format where data and protocol were merged, then one would first need to distinguish the protocol part of the test from the data part, so as to change the protocol while leaving the data intact, before the test can be applied at the SOC level. While this is possible, it can become a complicated process when protocol and data are not distinctly identified. However, if the core-level test exists in a format where data and protocol are separated, then the data portion of the test can be left intact and only the protocol would be changed to meet the SOC environment requirements. It follows that a pattern format where data and protocol are separated fits the embedded core test needs better than a format where data and protocol are merged. The process of manipulating test protocol at the SOC level constitutes pattern porting from the core-level to the SOC level and CTL, because of its requirement for separation between data and protocol, is uniquely suited for SOC level data porting.

In summary, the 1500 standard describes a DFT wrapper that surrounds a core to enable various types of testing of this core after it has been embedded into an SOC. This wrapper comprises a wrapper interface port, a Wrapper Instruction Register, a Wrapper Data Register, and a Wrapper BYpass register. The various 1500 wrapper components were designed with a significant level of flexibility that should prevent performance restrictions and allow a great variety of tests. These capabilities may not exist in legacy core test solutions. 1500 also provides a dual level of compliance that offers a much needed flexibility that takes into account the reality that efficient tests at the SOC level may require that some wrapper components are inserted by the core provider while others – that may affect test of SOC logic or affect quiescence at the SOC level during core test – are inserted by the core user. In addition, SOC level test flexibility needs, plug-and-play and test generation time handicaps related to large and complex SOCs can be alleviated by the 1500 architecture. This is facilitated by the use of CTL as the test pattern and test characteristics description language which allows for automation of test data porting and ease of test reuse at the SOC level.

Chapter 2
Why use the IEEE 1500 Standard?

2.1 Introduction

Before the question of "Why use the IEEE 1500 Standard?" is asked, another question should be addressed. Why do we need a wrapper at all? A wrapper is an isolation boundary between a core and the rest of the design. This isolation boundary allows test reuse to occur on the core as well as the capability to fully test the logic external to the core without having to exercise or have access to the core. Other advantages, for which the wrapper allows, are modular or partition-based debug, diagnosis and testing.

2.1.1 Test Reuse and Partitioning

When an SOC is created, it can be a massive effort to generate the test patterns for screening the device. This effort cannot begin until very late in the design flow as the SOC must be complete first. In an SOC that is composed of wrapped cores, this effort can be divided between different people and done at different times.

Each core is probably completed at a different time. If the delivered core is wrapped, an isolated set of test patterns can be generated for it at the time it is completed. These patterns are delivered with the core and can be post-processed to run at the SOC level. This partitioning of tasks means less work in the critical path at the end of the SOC design.

A hard core can also be delivered by another company. A wrapper allows the core to be tested with a minimum pinset. In addition, the wrapper can be used to test logic external to the core. This is especially important if the core user does not have access to the netlist of the hard core. A model of the wrapper can be created that does not infringe on any of the core IP, but it is enough information to read into a tool to create control and observe logic external to the core.

This parallel design strategy can shorten the design schedule. If a design change is needed in a core, it can be done locally. The timing and layout does not need to be redone for the entire design and test generation needs to be redone for the changed core only.

Consider the wrapped core and corresponding pattern in Figure 2. Note the port names on the core and in the pattern set. Once this core is embedded into an SOC, the core signals may change names, disappear or new signals may need to be added. The core and core pattern can be reused in many designs relieving some of the workload required for an SOC design.

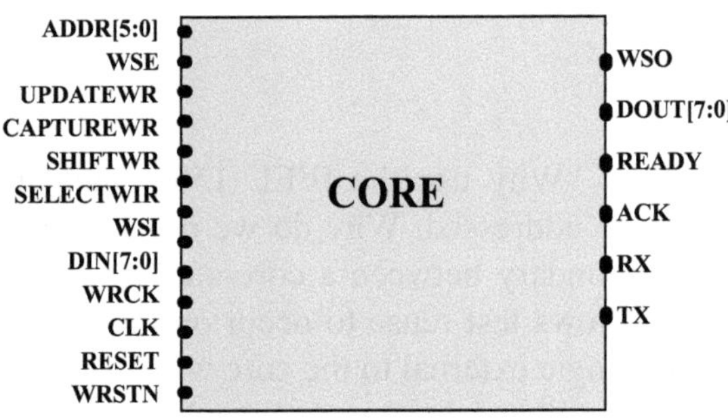

Core Pattern

```
WWUCSSWWCRW
RSPAHESRLES
CIDPILESKSO
K ATFE T E
  TUTC N T
  ERWT
  WERW
  RW I
   R R
1000000011X
1000010011X
1100110110X
1100110110X
1000110110X
10001101101
11001101101
11001101100
10001101100
```

Figure 2 Isolated Core with Pattern

As shown in Figure 2, the test pattern is limited to the test signals only. Constraining the functional signals to X or removing them during test pattern generation is common practice.

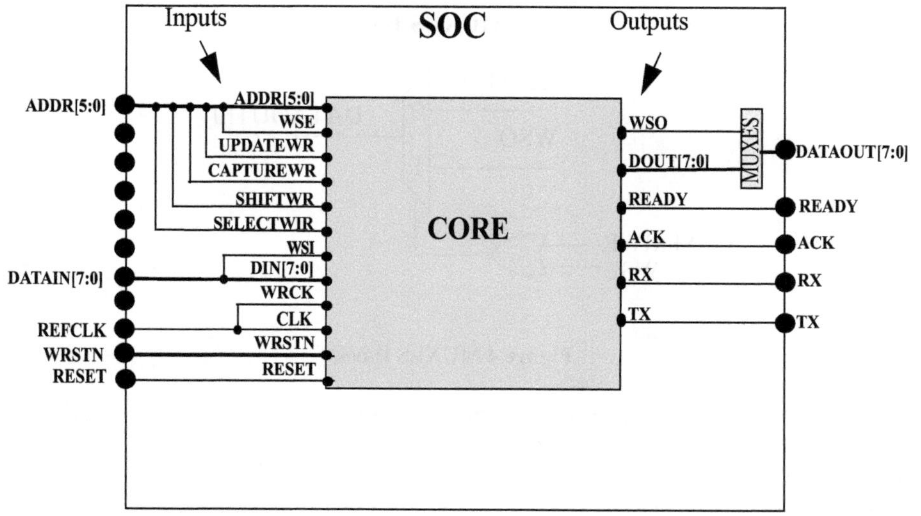

Figure 3 Core Embedded in an SOC

Figure 3 shows how the example wrapped core might be embedded into an SOC. All inputs are shown on the left side of the core and the outputs are shown on the right side of the core. Note that the SOC inputs, such as ADDR, can drive both the test inputs and the functional inputs on the core. These functional ports are not utilized during test as they are controlled and observed by the WBR. It is good practice to gate the test inputs off during functional mode - this is not shown in Figure 3. Also note that there is a multiplexer block (MUXES) connected to the WSO and DOUT outputs in Figure 3. The MUXES block determines which output ports of the core will be accessible from the DATAOUT pins. Figure 4 (page 16) shows what is inside the MUXES block. While the WSE or ShiftWR signal is high, the WSO signal is output to the DATAOUT[0] pin. While the WSE and ShiftWR signals are low, the DOUT[0] signal is output to the DATAOUT[0] pin. WSE and ShiftWR need to be held low (disabled) during functional mode – this is not shown in Figure 3.

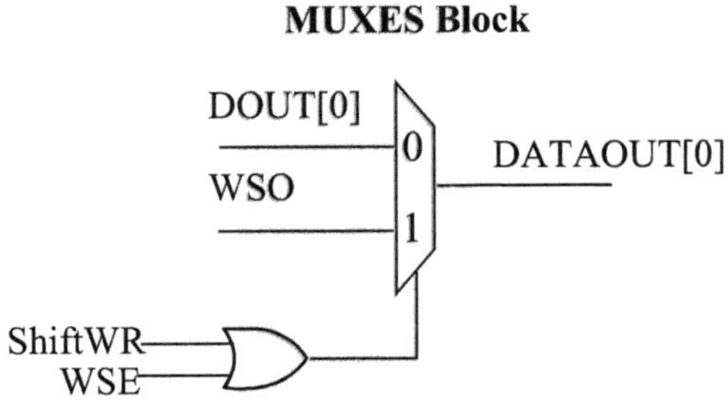

Figure 4 MUXES Block

Table 1 lists each core port and the SOC pin to which it is connected.

Table 1 Core Port to SOC Connection

Core Port	SOC Pin	Direction	Type
WRCK	REFCLK	Input	Test
WSI	DATAIN[0]	Input	Test
UPDATEWR	ADDR[0]	Input	Test
CAPTUREWR	ADDR[1]	Input	Test
SHIFTWR	ADDR[2]	Input	Test
SELECTWIR	ADDR[3]	Input	Test
WRSTN	WRSTN	Input	Test
WSE	ADDR[4]	Input	Test
DIN[7:0]	DATAIN[7:0]	Input	Func
ADDR[5:0]	ADDR[5:0]	Input	Func
READY	READY	Output	Func
CLK	REFCLK	Input	Func
RESET	RESET	Input	Func
WSO	DATAOUT[0]	Output	Test
DOUT[7:0]	DATAOUT[7:0]	Output	Func
ACK	ACK	Output	Func
RX	RX	Output	Func
TX	TX	Output	Func

If patterns are created for a hard core that does not have a wrapper, each signal in the pattern must be ported out to an external pin in order to control all of the inputs and observe all of the outputs. However, if the core is wrapped, only the dedicated test and minimal functional pins (e.g. clocks and asynchronous resets) need to be ported to the top level of the SOC.

After the core is embedded, the test pattern delivered with the core must be manipulated to be run from the SOC level. Figure 5 shows how the core pattern from Figure 2 (page 14) is converted so that it is able to run from the SOC level. The names of each signal has been changed to accommodate the connections of the core pins to the SOC level.

In other words though the data is the same, the protocol has changed as described in Section *1.3.1* (see page 9).

Core Pattern at SOC Level

```
R D A A A A A R D W
E A D D D D D E A R
F T D D D D D S T S
C A R R R R R E A T
L I [0][1][2][3][4] T O N
K N               U
  [0]             T
                  [0]

1000000 1x0
10000101x0
11001100x1
11001100x1
10001100x1
1000110011
1100110011
1100110001
1000110001
```

Figure 5 Core Pattern at SOC Level

2.1.2 Partition-Based Testing, Debug and Diagnosis

Test, debug and diagnosis of an SOC can be an overwhelming task. If the SOC is partitioned, this task becomes much more manageable. If a pattern fails on the tester and the SOC is not partitioned, it takes much work to figure out which part of the SOC caused that failure. On the other hand, if the SOC has been partitioned into wrapped cores and a pattern fails, the area of the failure can more easily be traced to a specific entity and no other wrapped core needs to be addressed during debug and diagnosis. Power during test can also be measured per core if the power domains are segmented properly. The wrappers also partition the logic external to the cores, so that if there is any area of issue in the external logic, it can also be found more easily.

2.2 Why Was a Standard Needed?

There are many advantages to standardization. There are no surprises in the implementation of the standard as tools can be created and the flow automated, literature is written to explain how best to use a standard and even improve on it.

Currently cores from multiple companies or even from different divisions within a company, can be utilized in an SOC. The cores could be a combination of hard and soft cores. A hard core is one whose library has been chosen, layout and routing is complete and all test structures are in place; in fact it is ready to go into a specific fabrication process. Hard cores users need only instantiate the hard core into an SOC of the same technology. A soft core is one that is delivered in behavioral language. The customer must do all timing, layout, routing and much of the test structure. However, the customer has the freedom to choose the technology, test methodology, etc. of this core. These cores may or may not have wrappers. In addition, there may be user-defined logic (UDL) external to the cores that must also be tested as shown in Figure 6 (page 19). The SOC designer must determine how to connect these entities up for functional use. In addition, the SOC designer must determine how to test all of the logic in the SOC. Each of the hard cores may have a different test methodology, which the SOC designer must understand and make work together. If the cores do have wrappers, it may be difficult to get core wrappers synchronized in order to test the UDL properly. The soft cores may need their test methodology developed by the SOC designer, as does the UDL. This is a lot of extra work to assign to the SOC designer and it may be a high risk flow due to the fact that the SOC designer may not be familiar with the cores. This flow may also take an excessive amount of time. If the interface for each of these cores was the same or similar and that interface allowed for testability of user-defined logic, this would make the SOC designers job much easier, more timely and less prone to mistakes. If this interface is standardized, it would allow for ease of automation of wrapper creation, manipulation of patterns, SOC test connections and test scheduling.

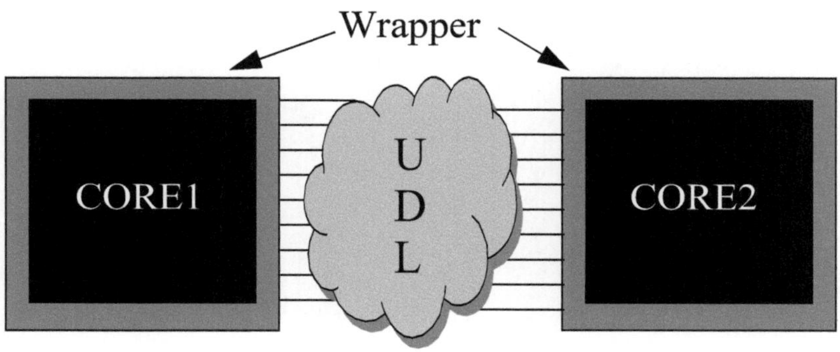

Figure 6 Simultaneous Access of Two Wrappers

The 1500 standard describes a standardized interface that will allow for all of the advantages listed above. This standard is rigid in some areas that require two or more cores' wrappers to be utilized in conjunction with each other. However, it is quite relaxed in other areas, such as the types of tests that can be allowed for testing cores or UDL. The standard lists a plethora of example wrapper cells to meet most test requirements. There is a standard mechanism for delivering instructions to all cores, a standard mechanism for bypassing wrapper boundary registers (WBRs) and a standard pin interface for use on the wrapper instruction register (WIR), wrapper bypass register (WBY) and WBR. It is a great combination of rules and flexibility that should allow automation of the creation of the wrappers, the stitching of test signals in an SOC and in conjunction with CTL, pattern manipulation for embedded cores and test scheduling.

Chapter 3
Illustration Example

3.1 Introduction

This section describes an example core with no wrapper (an unwrapped core). This example core will be used as a reference throughout this book when needed to show how to employ various parts of the standard (e.g. WIR, instructions, WBR). A 1500 wrapper will be built piece by piece for this example unwrapped core. From a compliance standpoint the goal of this exercise is to create a 1500 wrapped compliant core. The generation of test patterns for this core will be assumed to be done at the wrapper level to conform to the scenario of a core provider.

3.2 Unwrapped Core

The unwrapped core example has just enough pins so that it can be utilized by each 1500 subject in this book. It is not an example of any particular functional core. Only items pertaining to the test interface are described in the following chapters. The functionality of the core is not addressed here, just the pin interface and the test functions of the core.

The example core is called EX. EX is a full scan core with memory BIST. An illustration of the boundary of EX is shown in Figure 7.

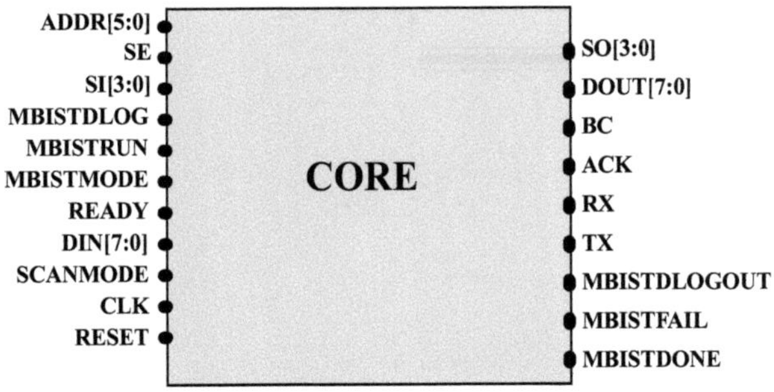

Figure 7 Unwrapped Core Example: EX

Table 2 shows the functionality of each signal. The test signals are in bold font. SCANMODE puts the core into the proper state to enable scan test of the core. All of the scan signals (SE, SI and SO) are gated off while SCANMODE is disabled (set to 0). MBISTMODE allows the core to test all of its memories with memory BIST (MBIST). All of the MBIST signals are gated off in functional mode while MBISTMODE is disabled (set to 0).

Each functional port is either registered or unregistered. A registered port has a functional register directly connected to the port of the core. A non-registered port has combinational logic between the register(s) and the port as shown in Figure 8. This information may be needed for the creation of the WBR.

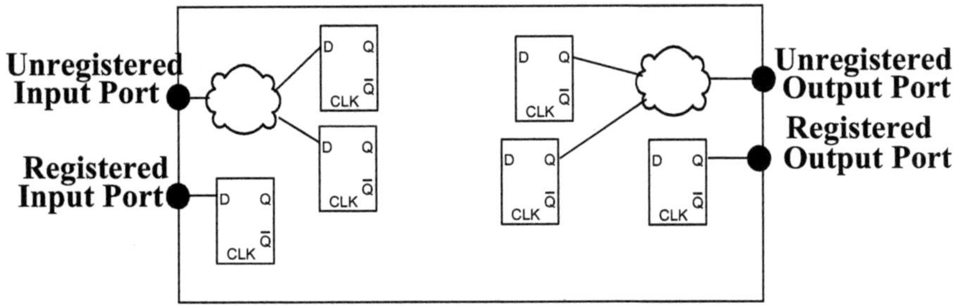

Figure 8 Example of Registered and Unregistered Ports

On the EX core, all of the functional inputs and outputs are registered, except for CLK, RESET and READY as shown in Table 2.

Table 2 UnWrapped Core Pins

Signal Name	*I/O*	*Function*	*Registered*
CLK	Input	Functional Clock	No
ACK	Output	Acknowledge Control Signal	Yes
RX	Output	Receive Signal	Yes
TX	Output	Transmit Signal	Yes
RESET	Input	Functional Core Asynchronous Reset	No
ADDR[5:0]	Input	Functional Address Pins	Yes
DIN[7:0]	Input	Functional Data Input Pins	Yes
DOUT[7:0]	Output	Functional Data Output Pins	Yes
READY	Input	Functional Ready Signal	No
BC	Output	Bus Control Signal	Yes
SCANMODE	Input	Puts the core into scan mode	
SE	Input	Scan Enable	
SI[3:0]	Input	Scan Input Bus	
SO[3:0]	Output	Scan Output Bus	
MBISTMODE	Input	Enables MBIST Mode	
MBISTDLOG	Input	MBIST Datalog Enable	
MBISTRUN	Input	Starts MBIST	
MBISTDLOGOUT	Output	MBIST Datalog Output	
MBISTDONE	Output	Indicates MBIST Complete	
MBISTFAIL	Output	Indicates MBIST Fail	

The description, thus far, may be enough information to create a WBR. More information is needed to create the Wrapper Instruction Register (WIR). The WIR controls the mode of the core (e.g. test mode, functional mode). Thus, the mode information is required.

The EX core has 3 different modes; functional mode, scan mode and MBIST mode. Scan mode is employed to scan test the logic, within the core, but external to the memories. While SCANMODE is enabled (high), scan mode is engaged. MBIST mode is utilized to test all of the memories within the core. While MBISTMODE is enabled (high) and SCANMODE is disabled (low), the memory tests are enabled. While MBISTMODE and SCANMODE are both disabled, functional mode is enabled as shown in Table 3.

Table 3 Core Modes

Mode	*SCANMODE*	*MBISTMODE*
Scan Mode	1	0
MBIST Mode	0	1
Functional Mode	0	0

During scan mode, the scan chains are accessed through the parallel scan ports SI[3:0] and SO[3:0]. These ports access four separate internal scan chains. Multiple scan chains can help reduce the vector set depth, which makes the test time significantly shorter. The WBY does not need information from the core to be created. It connects to the Wrapper Serial Port (WSP) only and does not interact with the core. The WSP is discussed in Section *4.1.1* (see page 29).

In order to determine which instructions are needed, information such as core modes is required. This information has been provided. However, to build the WIR, information from the WBR is also needed since the WIR controls the configuration of the WBR (e.g. internal test mode, external test mode). What types of configurations does the WBR have? Are there multiple WBR scan chains? Can there be a parallel external test instruction such as WP_EXTEST, defined in Section *5.1.2.1* (see page 84), in addition to the serial external test instruction (WS_EXTEST) defined in Section *5.1.1.2* (see page 80)?

The WIR needs more information than is provided by the unwrapped core. The number of instructions that are needed must be known so that the proper shift register can be built. Static test signal ports that control the configuration of the WBR must also be known.

3.3 Writing CTL For the EX Core

This section covers CTL for the EX core. This initial CTL code will apply to the EX core but will later be adapted to the EX wrapper. The EX wrapper CTL will be created progressively in the following chapters as we learn more about various components of the 1500 wrapper.

3.3.1 CTL Code Convention

CTL (just like STIL) is a structured language with nested blocks delimited by curly braces ({ }). Indentation will be used as much as possible to improve readability of the CTL code. The following 1450.6 convention will apply:

- Keywords reserved in CTL will be highlighted in bold
- The asterisk character (*) will indicate that the preceding keyword or group of keywords can be omitted or appear several times (same meaning as in regular expressions)
- The plus character (+) will indicate that the preceding keyword or group of keywords can appear once or several times (same meaning as in regular expressions)
- Parentheses will be used to delimit a group of keywords when necessary. Unless followed by an asterisk or a “plus” character, the parentheses will indicate that a keyword or group of keywords can be omitted or appear once.
- The “pipe” character (|) will serve as the “or” function to indicate that one choice is to be made within a group of possible keywords. The group begins with “<“ and ends with “>”
- Small cap text will indicate user data

3.3.2 CTL Code For the Unwrapped EX Core

The first step in writing CTL syntax is to declare boundary information about the design being described. This involves listing all the signals that exist at the design level and results in the following CTL code for the EX core.

```
<3.1>   STIL 1.0 {
<3.2>   Design 2005;
<3.3>   CTL 2005;
<3.4>   }
<3.5>
<3.6>   Signals {
<3.7>           CLK In;
<3.8>           RESET In;
<3.9>           ADDR[0..5] In;
<3.10>          DIN[0..7] In;
<3.11>          DOUT[0..7] Out;
<3.12>          READY In;
<3.13>          BC Out;
<3.14>          SCANMODE In;
<3.15>          SE In;
<3.16>          SI[0..3] In;
<3.17>          SO[0..3] Out;
<3.18>          MBISTDLOG In;
<3.19>          MBISTMODE In;
<3.20>          MBISTRUN In;
<3.21>          MBISTDLOGOUT Out;
<3.22>          MBISTDONE Out;
<3.23>          MBISTFAIL Out;
<3.24>          ACK Out;
<3.25>          RX Out;
<3.26>          TX Out;
<3.27>
<3.28>  }
<3.29>  SignalGroups {
<3.30>          all_non_mbist='CLK + RESET + ADDR[0..5] + DIN[0..7] +
                DOUT[0..7] + READY + BC + SCANMODE + SE + SI[0..3] +
                SO[0..3] + ACK + RX + TX';
<3.31>          all_mbist='MBISTDLOG + MBISTMODE + MBISTRUN +
                MBISTDLOGOUT + MBISTDONE + MBISTFAIL';
<3.32>  }
```

Lines 3.1 to 3.4 define that we are about to read STIL language and lists the characteristics of this STIL code. These characteristics are defined on lines 3.2 and 3.3. Line 3.2 specifies that the 1450.1 extension of STIL will be used and line 3.3 specifies that the CTL extension of the STIL language will also be used. These lines also provide the compliance year or version of these extensions, in our case 2005. Because CTL is based on both 1450 and 1450.1, lines 3.1 to 3.4 are going to begin every CTL code written to 1500 logic. This is because the 1500 standard specifically requires the 2005 version of CTL which in turn requires 1450.1.

Lines 3.6 to 3.28 declare the interface of the design for which the CTL code is being written. In our case these lines describe IOs of the EX core and their respective direction. This brings us to a key characteristic of CTL. The **Signals** block declares the IO terminals but does not specify the function of these terminals. In CTL the test function corresponding to each terminal is attached to a specific test mode. As a result a signal declared in the **Signals** block can serve as a scan input pin in a particular test mode but become a regular primary input pin in another test mode defined in the same CTL code. Due to the test mode-specific nature of CTL, 1500 specific characteristics will ultimately be attached to test modes as we will progressively discover throughout this book.

The **SignalGroups** block (lines 3.29 to 3.32) allows signals to be grouped so that the resulting group name can be referenced in other parts of the CTL code instead of referencing individual signal names comprising the group. EX core terminals are divided into two signal groups. The first terminal group is called "all_non_mbist" and includes all terminals with the exception of MBIST terminals. The MBIST terminals form the second terminal group called "all_mbist".

It should be noted that the section of CTL code discussed so far is STIL code. For our discussion, CTL can be viewed as STIL code where interface signals are defined, followed by one or more 1500 wrapper test modes which describe the function of each signal during each 1500 test mode. This structure is represented in Figure 9.

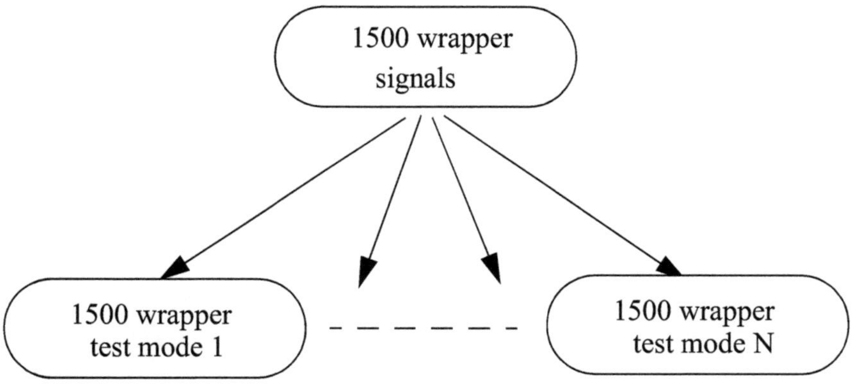

Figure 9 High-level CTL test mode structure

At this point, it is appropriate to clarify the definition of a test mode as this will be an important concept throughout this CTL writing exercise. A test mode can be defined as a specific configuration of control signals that set a design into a state where a specific type of tests can be applied to this design. With respect to our discussion this definition signifies that any particular configuration of the 1500 wrapper before and during application of test data can be identified as a test mode.

Chapter 4
Design of the IEEE 1500 Interface Port

The 1500 interface defines two types of access ports that anticipate serial, parallel and hybrid uses of a wrapped core. Each type of access (serial, parallel or hybrid) is characterized by the nature of the interface port that provides this access. A serial access is provided through the Wrapper Serial Port (WSP), a parallel access is provided through a Wrapper Parallel Port (WPP), and a hybrid access uses a combination of WSP and WPP signals. This classification of wrapper access type also extends to the 1500 standard instructions as these instructions are characterized by the type of access that they use to operate the wrapped core.

Besides defining design guidelines for the architecture of the wrapper interface signals, 1500 mandates the use of CTL for the description of these signals. In particular, timing information that ensures proper interfacing of the wrapper to its outside world is to be described in CTL.

In this section, the wrapper interface port is described as well as CTL requirements related to this port. In addition, the wrapper interface port creation process is depicted for the EX core.

4.1 Creation of the Wrapper Interface Port

4.1.1 Creation of the WSP

Provision of a WSP is mandatory on a 1500 wrapper. This requirement is meant to ensure a common and minimal architecture for all 1500 compliant wrappers. A primary benefit of this requirement is that it helps to achieve "Plug-and-Play" of 1500-compliant cores in the application of standard serial instructions. The WSP provides control signals to the wrapper as well as a single-bit data access to and from the wrapped core. Ten terminals (eight mandatory and two optional) comprise the WSP. In comparison with IEEE Std. 1149.1 (JTAG), the WSP terminal count is higher than the number of JTAG terminals but this is not an impediment as 1500 defines a core-level architecture whereas JTAG defines a chip-level architecture. This difference in purpose gives more flexibility to the design of the WSP and other 1500 interface port in terms of pin count requirements as creation of pins is less restrictive at the core-level than it is at the chip-level. The benefit is that the 1500 architecture does not need a state machine that would adapt a limited number of chip-level terminals to a larger number of protocols signals and states of a state machine. Instead, 1500 requires a block-level terminal for each of its test protocol needs. This has the very essential benefit of allowing seamless integration of 1500-compliant cores with a chip-level JTAG TAP controller by mapping WSP terminals to JTAG TAP states or protocol control signals. The WSP contains all terminals required to create this interface at the chip-level and therefore making the WSP mandatory guarantees that all 1500 compliant wrappers are designed with the flexibility to be integrated with a JTAG tap-controller at the chip-level if the SOC integrator chooses to do so. Section *11.2* (see page 261) discusses the interface between 1500 and 1149.1.

4.1.1.1 Accessing the 1500 Wrapper Through the WSP

The WSP contains dedicated terminals provided for the purpose of controlling the various wrapper components. These control signals form the Wrapper Serial Control (WSC) port. The WSC contains the following terminals:

- Wrapper clock terminal (WRCK)
- Wrapper reset terminal (WRSTN)
- Wrapper instruction register selection terminal (SelectWIR)
- Wrapper register shift control terminal (ShiftWR)
- Wrapper register capture control terminal (CaptureWR)
- Wrapper register update control terminal (UpdateWR)
- Wrapper data register transfer control terminal (TransferDR)
- Auxiliary wrapper clock terminal(s) (AUXCKn)

WSC terminals that provide protocol control such as the terminals providing the Shift, Capture, Update and Transfer control functions can be mapped to decoded JTAG state machine outputs. This allows WSC terminals to be sourced from JTAG circuitry.

In addition to the WSC the WSP contains two data terminals that provide serial data access to the 1500 wrapped core. These are:

- Wrapper Serial Input (WSI)
- Wrapper Serial output (WSO)

The following sections will help define the various 1500 wrapper interface terminals.

4.1.1.2 Understanding WRCK

The 1500 wrapper clock (WRCK) is a dedicated WSP terminal that provides clock function to all 1500 wrapper components. This function cannot be shared with any other wrapper-level clock terminal. Separation of the wrapper clock from functional clocks is essential to the 1500 architecture for the following reasons:

- **a.** a dedicated wrapper clock allows test wrapper operations (such as the preload operation) to be run concurrently with functional operation of the core
- **b.** a dedicated wrapper clock facilitates Plug-and-Play of 1500 wrappers by providing a single reference clock for the timing of all 1500 wrapper operations. Having a single reference clock also helps in the use of functional clocks in the wrapper, as the 1500 wrapper clock provides a single timing reference to which functional clocks can be synchronized.

Timing parameters associated with the WRCK signal are shown in Figure 10.

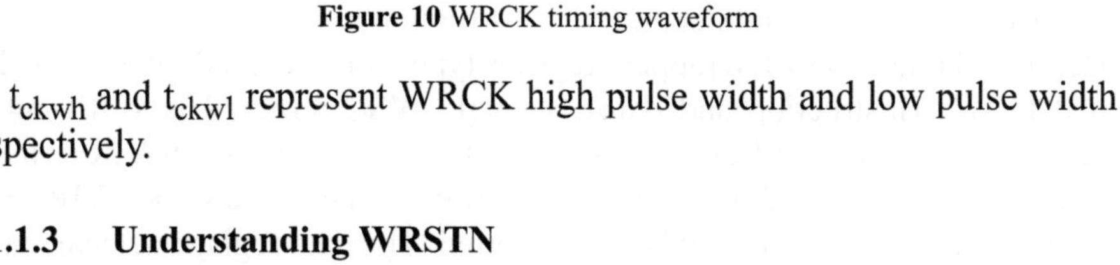

Figure 10 WRCK timing waveform

t_{ckwh} and t_{ckwl} represent WRCK high pulse width and low pulse width respectively.

4.1.1.3 Understanding WRSTN

The 1500 wrapper reset terminal (WRSTN) is similar to the JTAG wrapper reset function as WRSTN is active low. When WRSTN is active the wrapper is disabled. This puts the wrapped core into normal or mission mode. In addition, WRSTN has timing requirements with respect to the rising edge of WRCK. as represented in Figure 11. This is not so much to control when reset can be de-asserted than it is to determine when post-reset operations can begin.

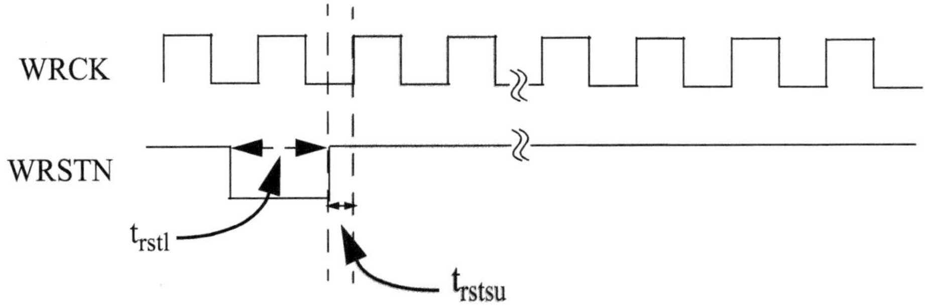

Figure 11 WRSTN timing waveform - asynchronous implementation

t_{rstsu} and t_{rstl} are Wrapper reset setup time and wrapper reset pulse width time respectively. These timing parameters were named by the standard without codification so as to allow a common usage without being restrictive as to the naming of these timing parameters. The standard prescribes an asynchronous behavior on WRSTN but also allows for an optional synchronous implementation of this signal. The timing waveform shown in Figure 11 represents the mandatory asynchronous implementation of WRSTN where only the de-assertion of WRSTN is timing critical. This timing information is meant to be defined in CTL so as to allow proper integration of the wrapped core at the SOC level.

4.1.1.4 Understanding SelectWIR

Each 1500 operation starts with the selection of the Wrapper Register (WR) that is the target of this operation. Two types of Wrapper Registers are defined by the 1500 architecture; the instruction register and a data register. Because the number of Wrapper Register types for the serial interface is limited to two (instruction and data), a single WSC terminal is sufficient for selection of a Wrapper Register type. This terminal is the SelectWIR terminal. The SelectWIR signal is active high. This signal selects the WIR when asserted and a data register when de-asserted. The data register selected at de-assertion of SelectWIR is determined by the instruction shifted and updated into the WIR while SelectWIR was asserted. Assertion of the SelectWIR terminal is expected to last as long as required to allow proper loading of the WIR. The timing waveform associated with the SelectWIR signal is depicted in Figure 12.

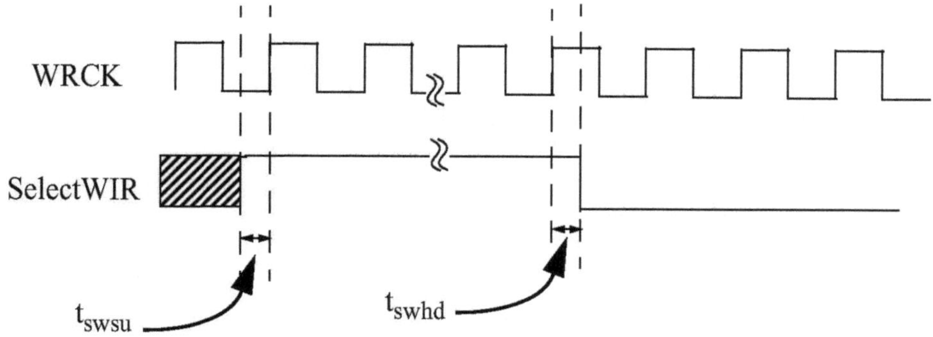

Figure 12 Timing waveform of the 1500 SelectWIR function

4.1.1.5 Understanding ShiftWR

The Shift operation of the 1500 wrapper is controlled by the ShiftWR (Shift Wrapper Register) terminal of the WSC. The assertion of this terminal establishes the main condition for a Shift operation on the 1500 Wrapper Registers. The timing waveform associated with the ShiftWR signal is represented in Figure 13.

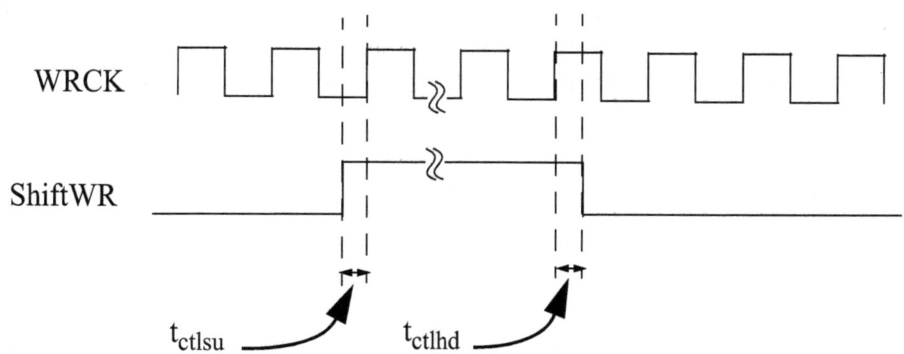

Figure 13 Timing waveform of the 1500 ShiftWR function

t_{ctlsu} and t_{ctlhd} represent setup and hold timing respectively.

4.1.1.6 Understanding CaptureWR

At the wrapper level, capture operation is controlled by the CaptureWR signal. When CaptureWR is asserted the capture operation occurs at the following timing-valid edge on wrapper storage elements that are meant to capture data. The timing waveform associated with the CaptureWR signal is shown in Figure 14.

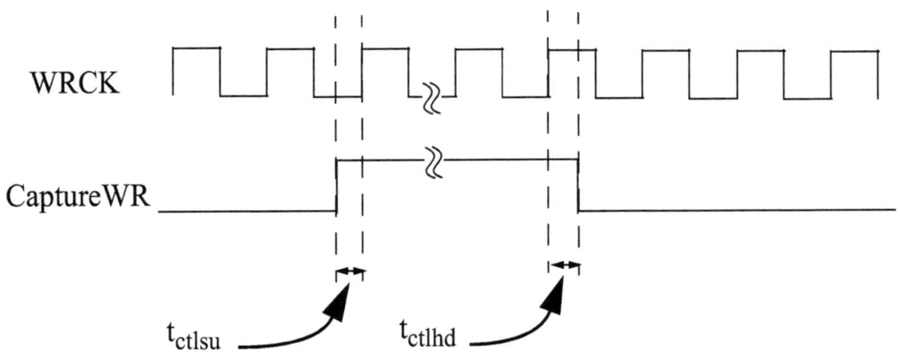

Figure 14 Timing waveform of the 1500 CaptureWR function

t_{ctlsu} and t_{ctlhd} represent setup and hold values respectively.

4.1.1.7 Understanding UpdateWR

The update operation is controlled by the UpdateWR signal and allows data present on shift paths to be loaded into update storage elements so as to preserve this data from destruction by a subsequent shift operation. This operation is similar to that defined in 1149.1. The timing waveform associated with the update operation is shown in Figure 15. The UpdateWR differs from other WSC signals in that it is sampled on the falling edge of WRCK.

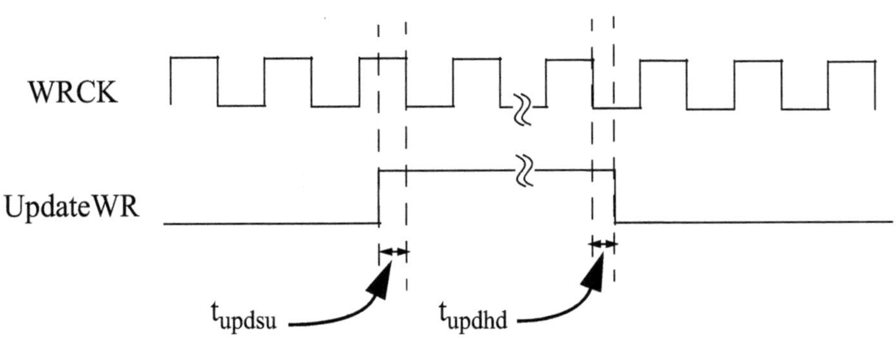

Figure 15 Timing waveform of the 1500 UpdateWR function

t_{updsu} and t_{updhd} represent setup and hold values.

4.1.1.8 Understanding TransferDR

Transfer is an optional operation that is controlled by the optional TransferDR wrapper interface signal. The transfer operation is specific to the WBR and will be further defined in Chapter 6. The timing waveform corresponding to this operation is shown in Figure 16.

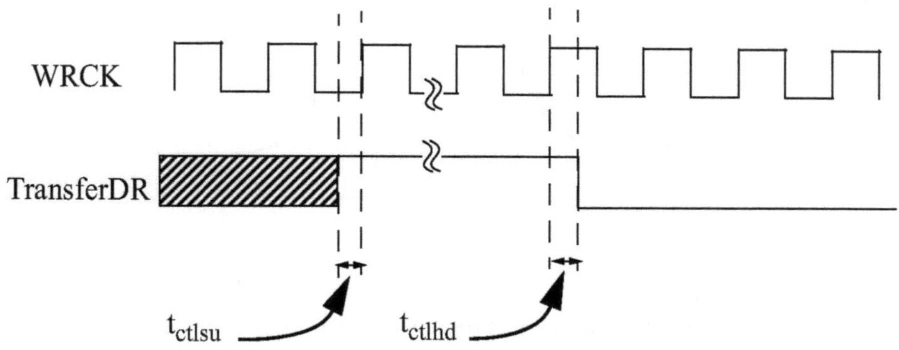

Figure 16 Timing waveform of the 1500 TransferDR function

t_{ctlsu} and t_{ctlhd} represent setup and hold values in Figure 16.

4.1.1.9 Understanding AUXCK Clocks

1500 anticipates the use of functional clocks (called auxiliary clocks) to control some components of the 1500 wrapper such as WBR cells. This permission to use non-dedicated 1500 clocks is meant solely to allow use of functional storage elements in 1500 compliant WBR cells or in user-defined data registers. More specifically, the WBY and WIR cannot be operated with functional clocks. In cases where an auxiliary clock is provided, WRCK must ultimately remain in control of the sequential operation of the WDR that is active during serial instructions. This implies that the auxiliary clock must be synchronized to WRCK. An example implementation described by the standard uses WRCK as an enable signal on a storage element that operates with an auxiliary clock. The example provided in the standard will be reused here for illustration purposes. The timing waveform of the AUXCK terminal is shown in Figure 17. The synchronization of AUXCK to WRCK results in the "Sync_AUXCK" signal which provides a one-to-one correspondence between a pulse on WRCK and a pulse on "Sync_AUXCK". It should be noted that there are many other ways to implement this synchronization and as a result many timing waveform descriptions are possible.

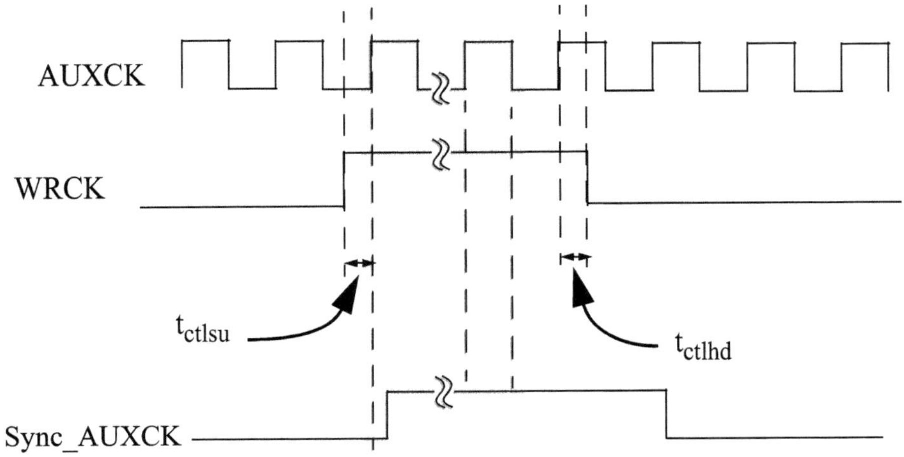

Figure 17 Timing waveform of the 1500 AUXCK terminal

t_{ctlsu} and t_{ctlhd} represent setup and hold values respectively.

4.1.1.10 Understanding the WSI Terminal

WSI is a mandatory single-bit data terminal that provides data to the data registers or the instruction register. The value of the SelectWIR terminal determines the destination (data register or instruction register) of data applied to WSI. The timing waveform of the WSI terminal is shown in Figure 18.

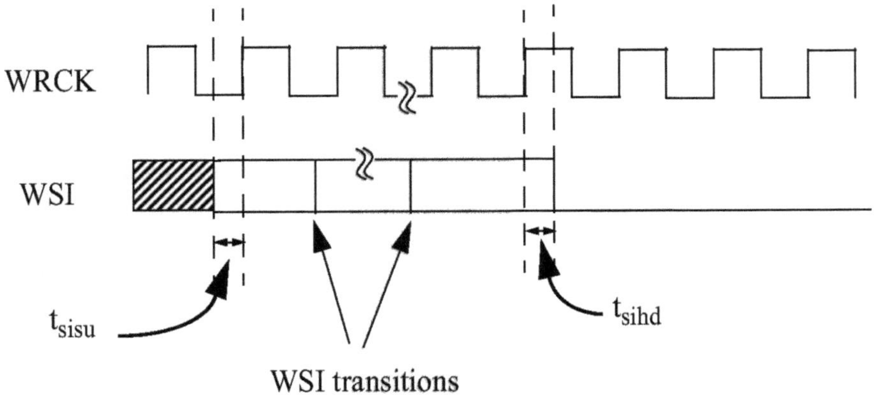

Figure 18 Timing waveform of the 1500 WSI terminal

t_{sisu} and t_{sihd} represent setup and hold values respectively.

4.1.1.11 Understanding the WSO Terminal

WSO is a mandatory single-bit data output terminal. This terminal is similar to WSI in that it is shared by 1500 data registers and the instruction register. Origin of the WSO data is determined by the value of the SelectWIR terminal. Unlike other WSP terminals, WSO changes on the falling edge of WRCK. This was done to facilitate interface timing requirements between two consecutive 1500 wrappers connected from WSO to WSI. Since WSI is sampled on the rising edge of WRCK, sampling WSO on the falling edge of WRCK allows data to meet WSI setup time on the wrapper serially connected to WSO. The timing waveform corresponding to the WSO terminal is shown in Figure 19.

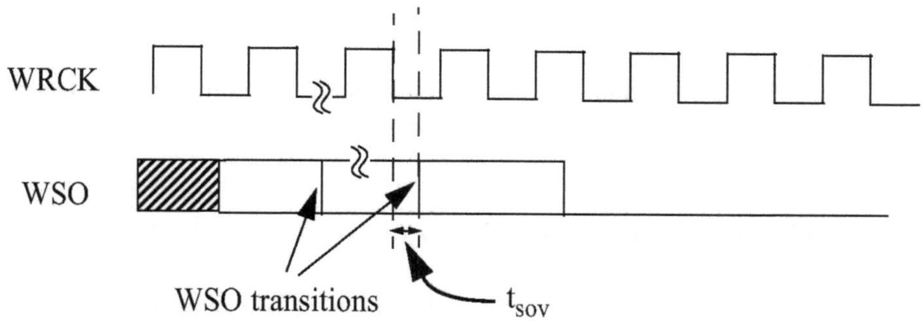

Figure 19 Timing waveform of the 1500 WSO terminal

t_{sov} represents the time is takes for data to be valid on WSO after the falling edge of WRCK.

4.1.2 Understanding the WPP

1500 mandates a serial wrapper implementation that uses the WSP so as to ensure plug-and-play of the corresponding mandatory instructions. However, some core designs have a requirement for data bandwidth that cannot be satisfied with a serial data interface to the wrapper. To satisfy this need, provisions were added into the standard to allow implementation of one or more parallel interfaces to the 1500 wrapper. Although the standard defines optional parallel and hybrid instructions to be used with the wrapper parallel port, very little was done to codify the nature of the parallel port itself. This gives control to the user of the standard in deciding what signals comprise the WPP so long as these signals, with the exception of WRCK and auxiliary clocks, are not shared between the WSP and the WPP. Without this exception, a test that uses a functional clock and WPP terminals might be forced to be called hybrid test if the functional clock is defined in the WSP as an auxiliary clock. It should be noted that WP_EXTEST, discussed in Section *5.1.2.1* (see page 84), is the only parallel instruction allowed to use WSC and WPP without being called a hybrid instruction. There is a strong distinction between WSP and WPP terminals and this was done so as to allow the serial data path of the wrapper to be used while the WPP is applying and observing a parallel test. In a case where multiple 1500 wrappers are connected together, this allows data to be shifted through a target wrapper to a downstream wrapper, while the target wrapper is operating a parallel test. In addition, WPP terminals are to be fully described in CTL if they are implemented, just like WSP terminals. The CTL keyword requirements for description of WPP terminals are the same as those discussed for the description of WSP terminals in Section *4.1.1* (see page 29).

Similar to the WSP, the WPP is formed with Wrapper Parallel Control (WPC) terminals and data terminals comprising Wrapper Parallel Input (WPI) and Wrapper Parallel Output (WPO).

Example WPP terminals will be created on the EX core to illustrate a possible usage for these user-defined terminals.

4.2 CTL Template for the Wrapper Interface Port

Defining CTL code for a 1500 wrapper interface port is equivalent to attaching a collection of 1500 test mode specific characteristics to signals that exist at the boundary of a wrapped core. Like STIL, CTL is made of nested block structures and the first question to be answered is, “where should this information go in the highly structured template of CTL?” To answer this question let’s remind ourselves of the high-level picture of our objective. Our main goal is to describe information at the boundary of a standalone design (in this case a wrapped-core). Therefore, we are interested in information that begins at the boundary of the wrapped core and continues inside the wrapped core. In other words we are mostly interested in defining the internal world of the wrapped-core design. In CTL, information describing the inside of a design goes into a block called **Internal**. This block is test mode specific because properties attached to the design boundary terminals can change from one test mode to another. This **Internal** block belongs to the following CTL code hierarchy structure:

```
STIL 1.0 { }
Signals { }
SignalGroups { }

Environment (DESIGN_NAME) {
        CTLMode (MODE_NAME){
        Internal { }
        }
}
```

The **Environment** block is specific to CTL and contains all CTL keywords defined by 1450.6 beyond the STIL syntax. This **Environment** block is a user-defined STIL code section where it is allowed to follow rules that extend native STIL mandates. "DESIGN_NAME" is optional and represents the name of the design being described in CTL. Inside the **Environment** block there will be a collection of test modes identified by the **CTLMode** block. The **CTLMode** keyword can accept an argument (MODE_NAME) that serves as name to the test mode being defined. However, the argument can be omitted and lead to what is referred to as a nameless test mode by 1450.6. An intended effect of a nameless **CTLMode** block is that this block is inherited by all other named **CTLMode** blocks. Therefore, we will purposefully not name the **CTLMode** block in this section so as to use this inheritance feature later in our efforts to write CTL code for a 1500 compliant wrapper. Blocks of CTL defined outside the test modes, but inside the **Environment** block, have to be brought into the scope of the test modes before their information can be used by these test modes. Unnamed CTL blocks defined outside the **Environment** block, like the **Signals** block, act as global entities. Their content is usable inside the **Environment** block.

Note: CTL examples in the 1500 standard document do not use the **CTLMode** keyword, but an older version of this keyword which was "**CTL**". This keyword was later changed from **CTL** to **CTLMode** by the 1450.6 working group. Defining the CTL language was not a 1500 working group charter and therefore this type of conflict is expected to be resolved by using the syntax defined in the approved 2005 version of 1450.6. In this case, this means using **CTLMode** to define CTL modes as opposed to using **CTL** to define these modes.

For the rest of our discussion it is important to characterize CTL as an extension of STIL, which builds upon keywords and constructs that exist in STIL without redefining them. As a result we will at times distinguish STIL keywords from CTL keywords in this book. This distinction mainly refers to the origin of these keywords and is artificial because ultimately all keywords used here are CTL keywords. However, because CTL is a very structured code, the distinction between STIL and CTL keywords is helpful in understanding the location of these keywords in the structure of the CTL code.

4.2.1 Structure of the CTL Internal block

In this section, the 1500-specific characteristics that are required in the **Internal** block are defined. From the 1500 standard perspective the **Internal** block comprises a collection of signals and their corresponding attributes. In general, these are wrapper interface port characteristics that need to be communicated from core provider to core user. In the process of defining the 1500 template for the **Internal** block, a generic signal referred to as "SIG_NAME" will be used. The template will therefore look as follows:

```
<4.10>  Environment{
<4.11>      CTLMode {
<4.12>          Internal {
<4.13>              SIG_NAME {
<4.14>              }
<4.15>          }
<4.16>      }
<4.17>  }
```

4.2.1.1 Specifying Electrical Characteristics Using the ElectricalProperty Keyword

The first requirement for the core-provider to core-user communication is that the information provided must be complete. In order to help achieve this, the 1500 standard chose to mandate that all design terminals pertaining to the level at which the CTL code is being written must be defined. The broad nature of this requirement encompasses not only all digital terminals but also all non-digital terminals. This signal-specific description is to be provided via use of the **ElectricalProperty** keyword of the CTL language. The corresponding CTL syntax is

ElectricalProperty
< Digital | Analog | Power | Ground > (Electrical_property_id));)

The "**Digital**" value of the argument is assumed by default if no other **ElectricalProperty** values are specified. This means that the **ElectricalProperty** keyword and its value may be entirely omitted in cases where the signal being described is a digital signal. The "**Electrical_property_id**" is an optional label used to identify a group of signals having the same electrical characteristics. The resulting CTL template is as follows:

```
<4.18>  Environment{
<4.19>      CTLMode {
<4.20>          Internal {
<4.21>              SIG_NAME {
<4.22>      ElectricalProperty < Digital | Analog |
        Power | Ground > Electrical_property_id;
<4.23>              } // SIG_NAME block
<4.24>          } // Internal block
<4.25>      } // CTLMode block
<4.26>  } // Environment block
```

4.2.1.2 Identifying 1500 Terminals Using the Wrapper Keyword

CTL is not a 1500 specific language and as such CTL can be used to describe a variety of test application schemes. But one of the characteristics of CTL is that it contains 1500 specific hooks built into the language. The **Wrapper** keyword (described below) defines one such hook.

There are two usages of the **Wrapper** keyword. The first usage is specific to the declaration of the wrapper interface port and the second one is specific to the identification of wrapper cells. These two usages are distinguished by the CTL block in which they appear. In this subsection we will focus on using the **Wrapper** keyword to define the interface port in the **Internal** CTL block.

From the 1500 perspective, the purpose of using the **Wrapper** keyword inside the **Internal** block is to distinguish 1500 serial interface terminals from other terminals. The CTL syntax of the **Internal** block **Wrapper** keyword follows:

(Wrapper <IEEE1500 | <u>**None**</u> | **User** USER_DEFINED > (**PinID** USER_DEFINED_PIN_ID);)

Where:

- A value of **IEEE1500** indicates that the signal being defined belongs to the 1500 serial interface port. In this case the PinID keyword must also be used to identify the specific 1500 terminal being declared. When defining 1500 terminals, the USER_DEFINED_PIN_ID value must be one of the WSP pin names defined by 1500. These are:

– WRCK
– WRSTN
– SelectWIR
– ShiftWR
– CaptureWR
– UpdateWR

- AUXCKn
- TransferDR
- WSI
- WSO

- A value of **None** is assumed by default when no other values are specified.
- A value of **"User** USER_DEFINED" identifies non-1500-WSP specific wrapper terminals such as WPP terminals. 1450.6 defines "USER_DEFINED" as an attribute that is pre-defined by a standard and that characterizes the **User** keyword. For our discussion we will use "WPP" as the value of "USER_DEFINED" when describing WPP terminals. The **PinID** value corresponding to a WPP terminal will be one of the 1500 WPP terminals (WPI, WPO), or the 1500 WPC terminal group. The **ElectricalProperty** value of "property" represents one of the arguments discussed in Section *4.2.1.1* (see page 40).

As allowed by CTL, in the following sections the **Wrapper** keyword will be omitted when there is no wrapper information to be specified. So far, we have defined the following CTL structure, where <value> is one of the 1500 terminal names described above:

```
<4.27>  Environment{
<4.28>      CTLMode {
<4.29>          Internal {
<4.30>              SIG_NAME {
<4.31>      ElectricalProperty < property> Electrical_property_id;
<4.32>      Wrapper (IEEE1500 | User WPP) PinID <value>;
<4.33>              } // SIG_NAME block
<4.34>          } // Internal block
<4.35>      } // CTLMode block
<4.36>  } // Environment block
```

4.2.1.3 Specifying Test Pin Type Using the DataType Keyword

The **DataType** keyword declares the test function associated with each signal being declared and the corresponding active state. The CTL syntax for the **DataType** keyword follows:

```
<4.37>  ( DataType ( data_type_enum )+;)
<4.38>  ( DataType ( data_type_enum )+ {
<4.39>      ( ActiveState < state> (Weak); )
<4.40>      ( AssumedInitialState < state > (Weak); )
<4.41>      ( ScanDataType ( ScanDataType_enum )+; )
<4.42>      ( DataRateForProtocol <Average|Maximum> INTEGER;)
<4.43>      ( ValueRange INTEGER INTEGER (CORE_INSTANCE_NAME)+; )*
<4.44>      ( UnusedRange INTEGER INTEGER;)*
<4.45>  } )* // DataType block
```

Where:

- data_type_enum specifies the signal type. Possible values are: *data_type_enum* = < **Asynchronous** | **Synchronous** | **In** | **Out** | **InOut** | **Constant** | **TestMode** | **Unused** | **UnusedDuringTest** | **Functional** | **TestControl** (*testcontrol_subtype_enum*)* | **TestData** (*testdata_subtype_enum*)* | **User** USER_DEFINED >
- testcontrol_subtype_enum = < **CaptureClock** | **CoreSelect** | **ClockEnable** | **InOutControl** | **Oscillator** | **OutDisable** | **OutEnable** | **MasterClock** | **MemoryRead**| **MemoryWrite** | **SlaveClock** | **Reset** | **ScanEnable** | **ScanMasterClock** | **ScanSlaveClock** | **TestAlgorithm** | **TestInterrupt** | **TestPortSelect** | **TestRun** | **TestWrapperControl** >
- testdata_subtype_enum = < **MemoryAddress** (**Row**|**Column**)*| **MemoryData**| **Indicator** (**TestDone** | **TestFail** | **TestInvalid**)* | **Regular** | **ScanDataIn** | **ScanDataOut**>

The **ActiveState** and **AssumeInitialState** keywords are identical in that both define values to be applied to the signal being defined e.g. **ForceDown, ForceUp** or values to be expected from the signal to be defined e.g. **ExpectHigh, ExpectOff.** As their names suggest, the difference between these two keywords is that the **AssumeInitialState** is to be used to specify initial state of the signal being defined at the start of every protocol in the **CTLMode** block, while **ActiveState** is to be used to specify the active state of the signal being defined. **ActiveState** and **AssumeInitialState** accept the same argument values. Possible **ActiveState** and **AssumeInitialState** arguments and their corresponding logic values are shown in Table 4.

Table 4 DataType ActiveState arguments and corresponding logic value mapping

ActiveState argument	*Logic value*
ForceDown	Logic 0
ForceUp	Logic 1
ForceOff	High impedance sate
ForceValid	Logic 1 or Logic 0
ExpectLow	Logic 0
ExpectHigh	Logic 1
ExpectOff	High impedance state
ExpectValid	Logic 0 or Logic 1

The **ScanDataType** keyword is to be used for scan input signals. From the 1500 perspective this keyword will be used on the scan inputs of the wrapper interface port.

Possible scan data type values are:

ScanDataType_enum = < **AddressGenerator** | **Boundary** | **Bypass** | **Counter** | **DataGenerator** | **Identification** | **Instruction** | **Internal** | **ResponseCompactor** | **User** USER_DEFINED >

The **DataType** keyword arguments are presented here for accuracy purpose, but not all of the CTL keywords will be defined here as doing so is beyond the scope of this book. Additional information on these keywords can be obtained in the 1450.6 document. Optional keywords that will not be used in this book will not be discussed further. So far, the following CTL structure has been defined:

```
<4.46>  Environment{
<4.47>      CTLMode {
<4.48>          Internal {
<4.49>              SIG_NAME {
<4.50>                  Wrapper (IEEE1500 | User WPP) PinID < value >;
<4.51>                  ( DataType ( <data_type_enum> )+ {
<4.52>                      ( ActiveState < value > (Weak); )
<4.53>                      ( AssumedInitialState < value > (Weak); )
<4.54>                      ( ScanDataType ( ScanDataType_enum )+; )
<4.55>                  } )* // DataType block
<4.56>              } // SIG_NAME block
<4.57>          } // Internal block
<4.58>      } // CTLMode block
<4.59>  } // Environment block
```

4.2.1.4 Specifying Input Pin Timing Using the DriveRequirements Keyword

In order to allow proper SOC integration of 1500-compliant cores, the 1500 standard mandates that timing information related to the wrapper interface port must be specified in CTL. This is to be done using the **DriveRequirements** CTL keyword. The **DriveRequirements** keyword syntax follows:

```
<4.60>  (DriveRequirements {
<4.61>      ( TimingNonSensitive; )
<4.62>      ( TimingSensitive {
<4.63>          ( Period < Min | Max > time_expr; )*
<4.64>          ( Pulse < High | Low > < Min | Max > time_expr; )*
<4.65>          ( Precision time_expr; )
<4.66>          ( EarliestTime time_expr; )
<4.67>          ( LatestTime time_expr; )
<4.68>          ( Reference sigref_expr {
<4.69>  ( SelfEdge < Leading | Trailing | LeadingTrailing > (INTEGER); )
<4.70>  ( ReferenceEdge < Leading | Trailing | LeadingTrailing > (INTEGER);)
<4.71>              ( Hold time_expr; )
<4.72>              ( Setup time_expr; )
<4.73>              ( Period real_expr;)
<4.74>              ( Precision time_expr; )
<4.75>          } )* // Reference block
<4.76>      (Waveform;)
<4.77>      } ) // TimingSensitive block
<4.78>  } ) // DriveRequirements block
```

Where:

- **TimingNonSensitive** characterizes a signal that is not timing critical. This keyword will not be used on mandatory 1500 wrapper terminals as these terminals are timing sensitive.
- The **TimingSensitive** block specifies timing information for timing-critical signals by using the following keywords:
  - * **Period**: minimum or maximum value of the signals period
  - * **Pulse**: minimum or maximum value of the signal's pulse width
  - * **Precision**: Uncertainty attached to the signal
  - * **EarliestTime**: minimum delay between the start of the test-cycle and the occurrence of a transition of the signal being described
  - * **LatestTime** minimum acceptable delay between the end of the test-cycle and the occurrence of a transition of the signal being defined
  - * The **Reference** keywords specify a signal from the CTL **Internal** block that serves as timing reference for the signal being defined in the **TimingSensitive** block. For our discussion the reference signal will be the 1500 wrapper clock. The timing relationship to the reference signal is further specified with the following keywords inside the **Reference** block:

    *__SelfEdge__: Defines the edge, of the signal being defined, that is timing critical. The syntax is **SelfEdge <Leading|Trailing|LeadingTrailing>** INTEGER
    where:
    Leading identifies the first edge of the signal from the beginning of the test cycle.
    Trailing identifies the edge that immediately follows the leading edge
    LeadingTrailing identifies either a Leading or a Trailing edge.
    INTEGER is used for signals that have a higher frequency than the test cycle, to indicate the edge number of the SelfEdge, counting from one.

    *__ReferenceEdge__: Specifies the edge, of the reference signal, that is used to define critical timing on the edge defined with **SelfEdge**. The syntax is similar to that of **SelfEdge** and has the

same argument definitions:
ReferenceEdge <**Leading|Trailing|LeadingTrailing**> INTEGER

***Hold**: Hold time requirement of **SelfEdge** with respect to **ReferenceEdge**

***Setup**: Setup time requirement of **SelfEdge** with respect to **ReferenceEdge**

***Period**: Specifies a period relationship between the signal being defined and its reference signal

***Precision**: Uncertainty on the values defined in the **Reference** block

* **Waveform**: Indicates that the timing characteristics of the signal being defined is to be extracted from the waveform table defined for this signal.

4.2.1.5 Specifying Output Pin Timing Using the StrobeRequirements Keyword

Output pin timing is to be specified using the **StrobeRequirements** CTL block. The syntax follows:

```
<4.79>  (StrobeRequirements {
<4.80>      ( TimingNonSensitive; )
<4.81>      ( TimingSensitive {
<4.82>          ( Precision time_expr; )
<4.83>          ( EarliestTimeValid time_expr; )
<4.84>          ( LatestTimeValid time_expr; )
<4.85>          ( EarliestChange time_expr; )
<4.86>          ( Reference sigref_expr {
<4.87>  ( SelfEdge < Leading | Trailing | LeadingTrailing > INTEGER; )
<4.88>  ( ReferenceEdge < Leading | Trailing | LeadingTrailing > INTEGER; )
<4.89>              ( EarliestTimeValid time_expr; )
<4.90>              ( LatestTimeValid time_expr; )
<4.91>              ( EarliestChange time_expr;)
<4.92>              ( Precision time_expr; )
<4.93>          } )* // Reference block
<4.94>      (Waveform;)
<4.95>      } ) // TimingSensitive block
<4.96>  } ) // StrobeRequirements block
```

The **StrobeRequirements** block uses the same keywords as the **DriveRequirements** keyword, with the exception of the following:

* **EarliestTimeValid**: minimum delay between the reference clock edge and the time the signal being described becomes valid.

- **LatestTimeValid**: Delay between the reference clock edge and the time the signal being described becomes invalid.
- **EarliestChange**: Earliest time (from the reference clock edge) where the signal being described is expected to change.

So far we have defined the following CTL structure:

```
<4.97>   Environment{
<4.98>       CTLMode {
<4.99>           Internal {
<4.100>              SIG_NAME {
<4.101>              Wrapper (IEEE1500 | User WPP) PinID < value >;
<4.102>                  (DataType ( data_type_enum )+ {
<4.103>                  (ActiveState < value > (Weak); )
<4.104>                  (AssumedInitialState < value > (Weak); )
<4.105>                  ( ScanDataType ( ScanDataType_enum )+; )
<4.106>                  } )* // DataType block
<4.107>
<4.108>              (DriveRequirements {
<4.109>                  ( TimingNonSensitive; )
<4.110>                  ( TimingSensitive {
<4.111>                      ( Period < Min | Max > time_expr; )*
<4.112>                      ( Pulse < High | Low > < Min | Max > time_expr; )*
<4.113>                      ( Precision time_expr; )
<4.114>                      ( EarliestTime time_expr; )
<4.115>                      ( LatestTime time_expr; )
<4.116>                      ( Reference sigref_expr {
<4.117>                          ( SelfEdge < Leading | Trailing |
                                     LeadingTrailing > (INTEGER); )
<4.118>                          ( ReferenceEdge < Leading | Trailing |
                                     LeadingTrailing > (INTEGER); )
<4.119>                          ( Hold time_expr; )
<4.120>                          ( Setup time_expr; )
<4.121>                          ( Period real_expr;)
<4.122>                          ( Precision time_expr; )
<4.123>                      } )* // Reference block
<4.124>                      (Waveform;)
<4.125>                  } ) // TimingSensitive block
<4.126>              } ) // DriveRequirements block
```

```
<4.127>                 (StrobeRequirements {
<4.128>                     ( TimingNonSensitive; )
<4.129>                     ( TimingSensitive {
<4.130>                         ( Precision time_expr; )
<4.131>                         ( EarliestTimeValid time_expr; )
<4.132>                         ( LatestTimeValid time_expr; )
<4.133>                         ( EarliestChange time_expr; )
<4.134>                         ( Reference sigref_expr {
<4.135>                             ( SelfEdge < Leading | Trailing |
                                        LeadingTrailing > INTEGER; )
<4.136>                             ( ReferenceEdge < Leading | Trailing |
                                        LeadingTrailing > INTEGER; )
<4.137>                             ( EarliestTimeValid time_expr; )
<4.138>                             ( LatestTimeValid time_expr; )
<4.139>                             ( EarliestChange time_expr;)
<4.140>                             ( Precision time_expr; )
<4.141>                         } )* // Reference block
<4.142>                         (Waveform;)
<4.143>                     } ) // TimingSensitive block
<4.144>                 } ) // StrobeRequirements block
<4.145>                 } // SIG_NAME block
<4.146>             } // Internal block
<4.147>         } // CTLMode block
<4.148>     } // Environment block
```

The above CTL structure defines a template that is sufficient for describing the 1500 wrapper interface port, but this template is not meant to be an exhaustive representation of 1450.6 capabilities. Instead, the template presents information that will allow description of 1500 compliant CTL code.

In the following sections we will use the above CTL template to achieve a 1500 compliant description of the wrapper interface port of the EX wrapper.

4.3 Wrapper Interface Port Creation for the EX Core

To begin the wrapper insertion process on the EX core, this core will be instantiated in a new design hierarchy and EX core terminals will be brought to the newly created upper design level. This new design level will be referred to as the wrapped core level. The instantiation process is not mandated by 1500 and as a result a single level of design hierarchy could contain the core and the 1500 wrapper components assuming that inside logic of the core is available during wrapper insertion.

For this discussion the following wrapper interface port terminals will be created:

- a 1500 WRCK terminal by the name of "WRCK"
- a 1500 WRSTN terminal by the name of "WRSTN"
- a 1500 SelectWIR terminal by the name of "SelectWIR"
- a 1500 ShiftWR terminal by the name of "ShiftWR"
- a 1500 CaptureWR terminal by the name of "CaptureWR"
- a 1500 UpdateWR terminal by the name of "UpdateWR"
- a 1500 WSI terminal by the name of "WSI"
- a 1500 WSO terminal by the name of "WSO"
- WPI terminals for scan test by the name of "WPSI[3:0]"
- WPO terminals for scan test by the name of "WPSO[3:0]"
- WPO terminals for MBIST with the following names: "MBISTDLOGOUT", "MBISTDONE", "MBISTFAIL"
- WPC terminals for scan test and MBIST with the following names: "WPSE", "MBISTRUN", "MBISTDLOG", "CLK", "WRCK".

The resulting wrapper level interface is shown in Figure 20.

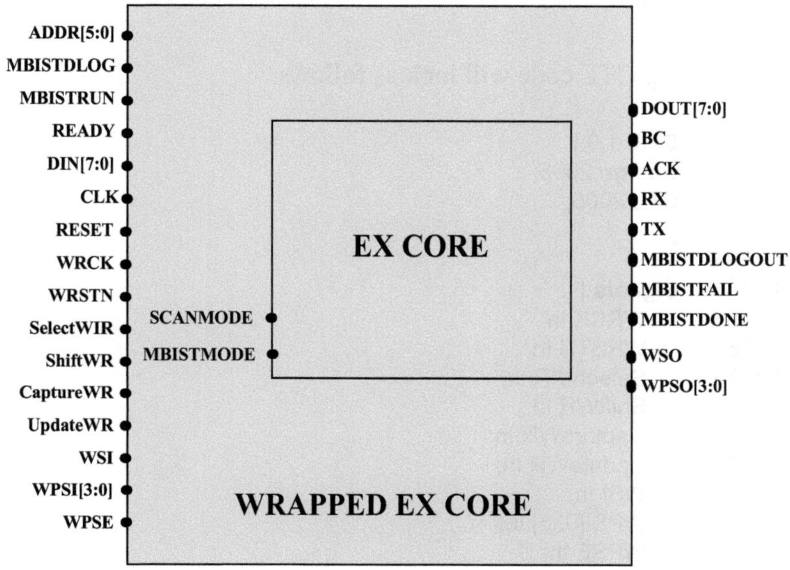

Figure 20 Wrapper design level for the EX core

It should be noted that the SCANMODE and MBISTMODE core terminals were not brought up to the wrapper level. These are "test mode" terminals and therefore need to be under control of the WIR per 1500 requirements. The corresponding EX core implementation is discussed in Section *8.3.3.4.2* (see page 210).

Now that the EX wrapper design level has been created, the **Signals** block defined in Section *3.3* (see page 23) at the EX core level can be ported to the EX wrapper level. The **SignalGroups** block will define groups of signals that will help subsequent CTL code writing steps. These groups are:

- "all_functional" which groups all functional EX wrapper terminals
- "wpp_mbist" which groups EX wrapper WPP terminals used during MBIST
- "wpp_scan" which groups EX wrapper terminals used during scan test
- "wpp_wrck" which represents an alias of WRCK. "wpp_wrck" will be used to apply CTL attributes to WRCK when the wrapper is in parallel mode

- "wsp_wrck" which represents an alias of WRCK. "wsp_wrck" will be used to apply CTL attributes to WRCK when the wrapper is in serial mode

The resulting CTL code will look as follows:

```
STIL 1.0 {
Design 2005;
CTL 2005;
}

Signals {
    WRCK In;
    WRSTN In;
    SelectWIR In;
    ShiftWR In;
    CaptureWR In;
    UpdateWR In;
    WSI In;
    WPSI[0..3] In;
    WPSE In;
    WSO Out;
    WPSO[0..3] Out;
    CLK In;
    RESET In;
    ADDR[0..5] In;
    DIN[0..7] In;
    DOUT[0..7] Out;
    READY In;
    BC Out;
    ACK Out;
    RX Out;
    TX Out;
    MBISTDLOG In;
    MBISTRUN In;
    MBISTDLOGOUT Out;
    MBISTDONE Out;
    MBISTFAIL Out;
}
SignalGroups {
    all_functional='CLK + RESET + ADDR[0..5] + DIN[0..7] +
        DOUT[0..7] + READY + BC + ACK + RX + TX';
    wpp_mbist='MBISTDLOG + MBISTRUN +
    MBISTDLOGOUT + MBISTDONE + MBISTFAIL + CLK';
    wpp_scan='WPSI[0..3] + WPSO[0..3] + WPSE + CLK';
    wpp_wrck='WRCK';
    wsp_wrck='WRCK';
}
```

The following sections will demonstrate CTL code creation for the EX wrapper terminals.

4.3.1 Declaring WRCK for the EX wrapper in serial mode

The 1500 WRCK terminal is expected to be identified in the CTL provided for the wrapped core. Per the template discussed in Section *4.2.1.2* (see page 41), this identification will look as follows for the wrapped EX core.

```
<4.189>   Environment EX_wrapper {
<4.190>      CTLMode {
<4.191>         Internal {
<4.192>            wsp_wrck {
<4.193>            DataType TestControl ScanMasterClock;
<4.194>            Wrapper IEEE1500 PinID WRCK;
<4.195>            } // wsp_wrck block
<4.196>         } // Internal block
<4.197>      }// CTLMode block
<4.198>   } // Environment block
```

The above CTL code defines the 1500 WRCK attribute on the EX wrapper WRCK terminal (of which "wsp_wrck" is an alias). CTL code line 4.192 represents the design terminal (or signal group) selected by the user to be used as 1500 WRCK. This is where an automation tool will look to identify WRCK in the CTL provided for the wrapped core. Next we will specify timing characteristics of the WRCK terminal using the **DriveRequirements** keyword defined in Section *4.2.1.4* (see page 45), along with the t_{ckwh} and t_{ckwl} timing parameters defined in Figure 10 (page 31). We will also specify a value for the clock period, as the sum of t_{ckwh} and t_{ckwl}. The resulting CTL including both the **Wrapper** and **DriveRequirements** specification follows:

```
<4.199>   Environment EX_wrapper {
<4.200>      CTLMode {
<4.201>         Internal {
<4.202>            wsp_wrck {
<4.203>               DataType TestControl ScanMasterClock;
<4.204>               Wrapper IEEE1500 PinID WRCK;
```

```
                DriveRequirements {
                    TimingSensitive {
                    Pulse High Min <tckwh>;
                    Pulse Low Min <tckwl>;
                    Period Min <tckwh + tckwl>;
                    } // TimingSensitive block
                } // DriveRequirements block
            } // wsp_wrck block
        } // Internal block
    } // CTLMode block
} // Environment block
```

4.3.2 Declaring WRSTN for the EX wrapper

Per the syntax defined in Section *4.2.1.2* (see page 41), the CTL code declaring the 1500 WRSTN terminal will look as follows:

```
WRSTN {
    DataType TestControl TestWrapperControl {
        ActiveState ForceDown;
    }
Wrapper IEEE1500 PinID WRSTN;
}
```

This syntax gives the EX wrapper WRSTN terminal the 1500 WRSTN attribute. We will also use the **DriveRequirements** keyword to specify setup and pulse width for the WRSTN terminal. Per the syntax discussed in Section *4.2.1.4* (see page 45) the updated CTL code including both the **Wrapper** and the **DriveRequirements** keywords will look as follows:

```
Environment EX_wrapper {
    CTLMode {
        Internal {
            WRSTN {
                DataType TestControl TestWrapperControl {
                ActiveState ForceDown;
                } // DataType block
                Wrapper IEEE1500 PinID WRSTN;
```

```
                    DriveRequirements {
                        TimingSensitive {
                            Pulse Low Min <t_rstl>;
                            Reference WRCK {
                                SelfEdge Trailing;
                                ReferenceEdge Leading;
                                Setup <t_rstsu>;
                            } // Reference block
                        } // TimingSensitive block
                    } // DriveRequirements block
                } // WRSTN block
            } // Internal block
        } // CTLMode block
    } // Environment block
```

4.3.3 Declaring SelectWIR for the EX Wrapper

The CTL code corresponding the SelectWIR terminal uses the same keywords introduced and used in the above sections along with the timing defined in Figure 12 (page 33). The CTL follows:

```
Environment EX_wrapper {
    CTLMode {
        Internal {
            SelectWIR {
                DataType TestControl TestWrapperControl {
                    ActiveState ForceUp;
                }
                Wrapper IEEE1500 PinID SelectWIR;
                DriveRequirements {
                    TimingSensitive {
                        Reference WRCK {
                            SelfEdge LeadingTrailing;
                            ReferenceEdge Leading;
                            Setup <t_swsu>;
                            Hold <t_swhd>;
                        } // Reference block
                    } // TimingSensitive block
                } // DriveRequirements block
            } // SelectWIR block
        } // Internal block
    } // CTLMode block
} // Environment
```

4.3.4 Declaring ShiftWR for the EX Wrapper

The EX wrapper uses a terminal by the name of "ShiftWR" to serve as 1500 ShiftWR signal. The corresponding ShiftWR CTL code follows:

```
Environment EX_wrapper {
   CTLMode {
      Internal {
         ShiftWR {
            DataType TestControl ScanEnable;
            DataType TestControl TestWrapperControl {
                     ActiveState ForceUp;
            }
            Wrapper IEEE1500 PinID ShiftWR;
            DriveRequirements {
               TimingSensitive {
                  Reference WRCK {
                     SelfEdge LeadingTrailing;
                     ReferenceEdge Leading;
                     Setup <t_ctlsu>;
                     Hold <t_ctlhd>;
                  } // Reference block
               } // TimingSensitive block
            } // DriveRequirements block
         } // ShiftWR block
      } // Internal block
   } // CTLMode block
} // Environment
```

t_{ctlsu} and t_{ctlhd} are setup and hold values corresponding to Figure 13 (page 33).

4.3.5 Declaring CaptureWR for the EX wrapper

The CTL syntax for describing CaptureWR uses the same syntax at that of the wrapper terminals described above. The resulting CTL code looks as follows.

```
Environment EX_wrapper {
   CTLMode {
      Internal {
         CaptureWR {
            DataType TestControl TestWrapperControl {
               ActiveState ForceUp;
            }
            Wrapper IEEE1500 PinID CaptureWR;
```

```
                    DriveRequirements {
                        TimingSensitive {
                            Reference WRCK {
                                SelfEdge LeadingTrailing;
                                ReferenceEdge Leading;
                                Setup <t_ctlsu>;
                                Hold <t_ctlhd>;
                            } // Reference block
                        } // TimingSensitive block
                    } // DriveRequirements block
                } // CaptureWR block
            } // Internal block
        } // CTLMode
    } // Environment
```

t_{ctlsu} and t_{ctlhd} are setup and hold values corresponding to Figure 14 (page 34).

4.3.6 Declaring UpdateWR for the EX Wrapper

The EX wrapper terminal corresponding to the 1500 Update signal is UpdateWR. Because timing parameters associated with the update operation are defined with respect to the falling edge of WRCK, the CTL **DriveRequirements** section corresponding to UpdateWR will define a **ReferenceEdge** value of "trailing".

```
Environment EX_wrapper {
    CTLMode {
        Internal {
            UpdateWR {
                DataType TestControl TestWrapperControl {
                    ActiveState ForceUp;
                }
                Wrapper IEEE1500 PinID UpdateWR;
                DriveRequirements {
                    TimingSensitive {
                        Reference WRCK {
                            SelfEdge LeadingTrailing;
                            ReferenceEdge Trailing;
                            Setup <t_updsu>;
                            Hold <t_updhd>;
                        } // Reference block
                    } // TimingSensitive block
                } // DriveRequirements block
            }// UpdateWR block
        } // Internal block
    } // CTLMode block
} // Environment block
```

t_{updsu} and t_{updhd} are setup and hold values corresponding to Figure 15 (page 34).

4.3.7 Declaring WSI for the EX Wrapper

The following represents description of the WSI terminal in CTL using the same CTL constructs as previous WSP CTL descriptions. The argument of the **SelfEdge** keyword will change to **LeadingTrailing** in order to accommodate both types of data transition on WSI.

```
Environment EX_wrapper {
    CTLMode {
        Internal {
            WSI {
                DataType Testdata ScanDataIn {
                    ScanDataType Boundary;
                }
                Wrapper IEEE1500 PinID WSI;
                DriveRequirements {
                    TimingSensitive {
                        Reference WRCK {
                            SelfEdge LeadingTrailing;
                            ReferenceEdge Leading;
                            Setup <t_sisu>;
                            Hold <t_sihd>;
                        } // Reference block
                    } // TimingSensitive block
                } // DriveRequirements block
            } // WSI block
        } // Internal block
    } // CTLMode block
} // Environment block
```

t_{sisu} and t_{sihd} are setup and hold values corresponding to Figure 18 (page 36).

4.3.8 Declaring WSO for the EX Wrapper

The following is a description of the WSO terminal in CTL. The main difference between the following WSO CTL and previous WSP CTL descriptions is that we will use the **StrobeRequirements** construct to specify WSO timing because we are describing an output terminal. The resulting CTL description follows.

```
Environment EX_wrapper {
    CTLMode {
        Internal {
            WSO {
                DataType TestData ScanDataOut {
                    ScanDataType Boundary;
                }
                Wrapper IEEE1500 PinID WSO;
                StrobeRequirements {
                    TimingSensitive {
                        Reference WRCK {
                            SelfEdge LeadingTrailing;
                            ReferenceEdge Trailing;
                            EarliestTimeValid <t_sov>;
                        } // Reference block
                    } // TimingSensitive block
                } // StrobeRequirements block
            } // WSO block
        } // Internal block
    } // CTLMode block
} // Environment block
```

t_{sov} represents the valid output time corresponding to Figure 19 (page 37).

WPP terminals will be added to the EX wrapper and used during test to provide more data bandwidth than the default serial interface. The test using the WPP will make use of a scan protocol and will require a scan enable terminal in addition to the WPSI, WPSO pair. The scan enable terminal will be called WPSE. For timing considerations we will consider that WPSI and WPSO share the same timing parameters as WSI and WSO.

4.3.9 Declaring WPC for the EX Wrapper

The WPC will be composed of WRCK, CLK (the EX wrapper functional clock), and three wrapper signals serving as WPC terminals. These three signals are: WPSE for scan mode control, and MBISTRUN and MBISTLOG for MBIST mode control.

4.3.9.1 Declaring WRCK for the EX Wrapper in parallel mode

Although WRCK is a WSC terminal, it is allowed to use this terminal in the WPC as well. The corresponding CTL code follows.

```
Environment EX_wrapper {
    CTLMode {
        Internal {
            wpp_wrck {
                DataType TestControl ScanMasterClock;
                Wrapper User WPP PinID WPC;
                DriveRequirements {
                    TimingSensitive {
                        Pulse High Min <tckwh>;
                        Pulse Low Min <tckwl>;
                        Period Min <tckwh + tckwl>;
                    } // TimingSensitive block
                } // DriveRequirements block
            } // wpp_wrck block
        } // Internal block
    } // CTLMode block
} // Environment block
```

The t_{ckwh} and t_{ckwl} timing parameters are defined in Figure 10 (page 31).

4.3.9.2 Declaring a Functional Clock for the EX Wrapper

The EX wrapper has a functional clock called “CLK”. Although some of the EX wrapper WBR cells will reuse functional storage elements, bypass logic will be added to the clock circuitry so these cells remain under the control of WRCK during serial test. This eliminates the need to declare “CLK” as a WSC auxiliary clock because the EX wrapper serial test will not require the use of “CLK”. However, EX wrapper parallel instructions will require “CLK” as they operate functional storage elements inside the core. For this reason “CLK” will be defined as WPC terminal, given that the standard provides large amplitude regarding the creation of the WPP. The corresponding “CLK” CTL code follows, assuming the timing parameters defined in Figure 17 (page 36).

```
Environment EX_wrapper {
    CTLMode {
        Internal {
            CLK {
                DataType TestControl ScanMasterClock;
                Wrapper User WPP PinID WPC;
```

```
                        DriveRequirements {
                            TimingSensitive {
                                Reference WRCK {
                                    SelfEdge Leading;
                                    ReferenceEdge Leading;
                                    Setup <t_ctlsu>;
                                    Hold <t_ctlhd>;
                                } // Reference block
                            } // TimingSensitive block
                        } // DriveRequirements block
                    } // CLK block
                } // Internal block
            } // CTLMode block
        } // Environment
```

t_{ctlsu} and t_{ctlhd} are setup and hold values corresponding to Figure 17 (page 36).

4.3.9.3 Declaring WPC for the EX wrapper scan mode

In scan mode WPSE serves as a scan enable signal that follows a traditional scan protocol. For our discussion we will assume that the WPSE terminals has the same timing as ShiftWR. This timing is shown in Figure 21.

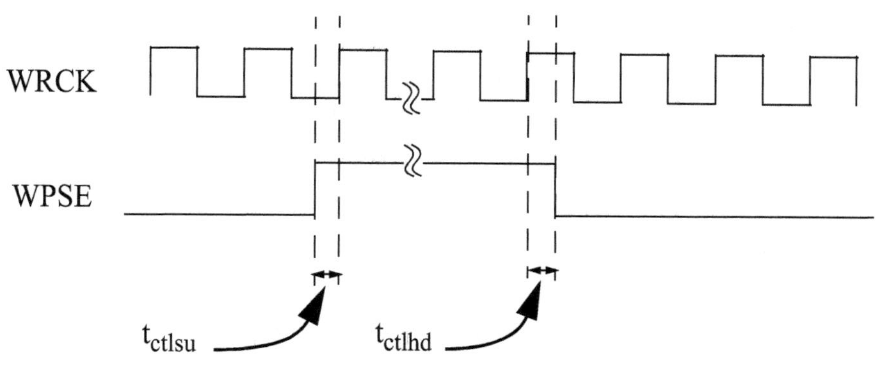

Figure 21 Timing waveform of the EX wrapper WPSE terminal

t_{ctlsu} and t_{ctlhd} represent setup and hold values respectively.
The CTL code corresponding to WPSE follows.

```
Environment EX_wrapper {
    CTLMode {
        Internal {
            WPSE {
                DataType TestControl TestWrapperControl;
```

```
                DataType TestControl ScanEnable{
                    ActiveState ForceValid;
                }
                Wrapper User WPP PinID WPC;
                DriveRequirements {
                    TimingSensitive {
                        Reference WRCK {
                            SelfEdge LeadingTrailing;
                            ReferenceEdge Leading;
                            Setup <tctlsu>;
                            Hold <tctlhd>;
                        } // Reference block
                    } // TimingSensitive block
                } // DriveRequirements block
            } // WPSE block
        } // Internal block
    } // CTLMode block
} // Environment block
```

t_{ctlsu} and t_{ctlhd} represent setup and hold values corresponding to Figure 21.

4.3.9.4 Declaring WPC for the EX wrapper MBIST mode

In MBIST mode two WPC control terminals named MBISTRUN and MBISTLOG will be used to control the test. The corresponding timing is shown in Figure 22.

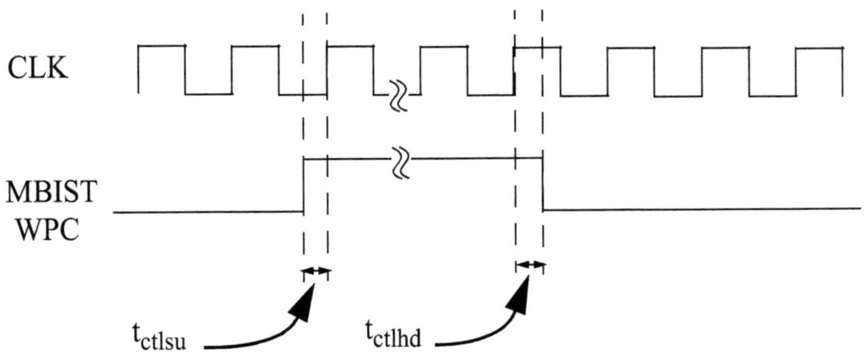

Figure 22 Timing waveform of the EX wrapper MBIST WPC terminals

t_{ctlsu} and t_{ctlhd} represent setup and hold values respectively.
The CTL code corresponding to the MBIST mode WPC terminals follows.

```
Environment EX_wrapper {
    CTLMode {
        Internal {
            MBISTRUN {
                DataType TestControl TestWrapperControl;
                Wrapper User WPP PinID WPC;
                DriveRequirements {
                    TimingSensitive {
                        Reference CLK {
                            SelfEdge LeadingTrailing;
                            ReferenceEdge Leading;
                            Setup <t_ctlsu>;
                            Hold <t_ctlhd>;
                        } // Reference block
                    } // TimingSensitive block
                } // DriveRequirements block
            } // MBISTRUN block
            MBISTLOG {
                DataType TestControl TestWrapperControl;
                Wrapper User WPP PinID WPC;
                DriveRequirements {
                    TimingSensitive {
                        Reference CLK {
                            SelfEdge LeadingTrailing;
                            ReferenceEdge Leading;
                            Setup <t_ctlsu>;
                            Hold <t_ctlhd>;
                        } // Reference block
                    } // TimingSensitive block
                } // DriveRequirements block
            } // MBISTLOG block
        } // Internal block
    } // CTLMode block
} // Environment block
```

t_{ctlsu} and t_{ctlhd} represent setup and hold values corresponding to Figure 22 (page 62).

4.3.10 Declaring WPI for the EX Wrapper

For the EX wrapper, WPI terminals will share the same timing as WSI, since these terminals have the same (scan input) function. This timing is shown in Figure 23 (page 64). All CTL code examples shown so far have EX wrapper terminal names that are identical to the CTL **PinID** argument of the **Wrapper** keyword e.g. the CTL **PinID** argument of the EX wrapper "WSI" terminal is "WSI". However, maintaining this match between terminal names and the corresponding CTL **PinID** argument is not a requirement. This will be illustrated with the WPP terminals, as these terminals have EX wrapper names that are different from the 1500 interface pin names (e.g. WPSI vs. WPI).

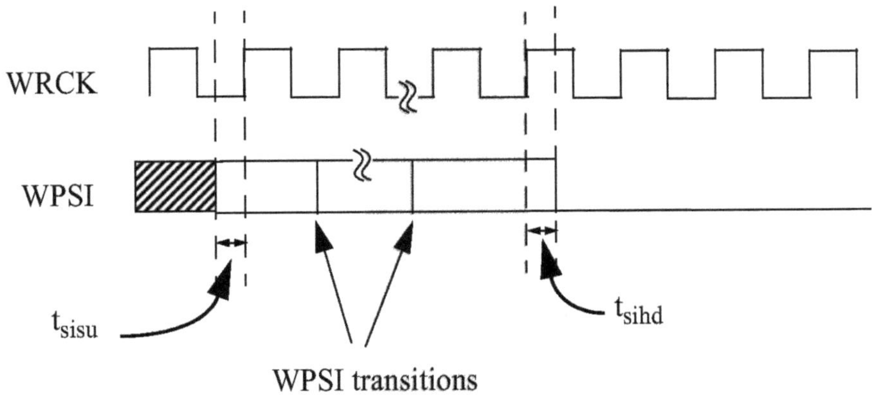

Figure 23 Timing waveform of the WPSI terminal

t_{sisu} and t_{sihd} represent setup and hold values respectively.

Four WPI terminals will be created for the EX wrapper in scan mode and named WPSI[0..3]. The resulting CTL code follows.

```
Environment EX_wrapper {
    CTLMode {
        Internal {
            WPSI [0..3] {
                DataType Testdata ScanDataIn {
                    ScanDataType Boundary;
                }
                Wrapper User WPP PinID WPI;
                DriveRequirements {
                    TimingSensitive {
                        Reference WRCK {
                            SelfEdge LeadingTrailing;
                            ReferenceEdge Leading;
                            Setup <tsisu>;
                            Hold <tsihd>;
                        } // Reference block
                    } ) // TimingSensitive block
                } ) // DriveRequirements block
            } // WPSI [0..3] block
        } // Internal block
    } // CTLMode block
} // Environment block
```

t_{sisu} and t_{sihd} are setup and hold values corresponding to Figure 23.

4.3.11 Declaring WPO for the EX Wrapper

Seven WPO terminals will be created on the EX wrapper (four terminals for scan test and three for MBIST).

4.3.11.1 Declaring WPO for the EX wrapper scan test mode

WPSO[3:0] will serve as WPO terminals in scan test mode. These terminals will have the same timing as the WSO terminal. In particular, this implies that WPSO will be sampled on the falling edge of WRCK, as shown in Figure 24. This is not a 1500 mandate, on WPO terminals, but a conservative approach that should ease timing considerations at the interface between the EX wrapper and its SOC host during parallel instructions.

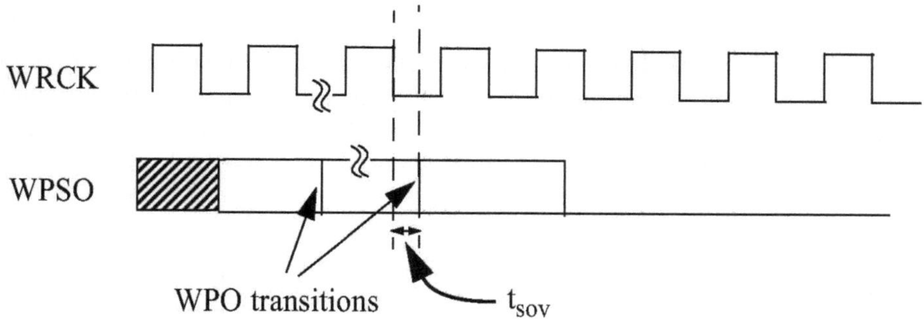

Figure 24 Timing waveform of the WPSO terminal

t_{sov} represents the time is takes for data to be valid on WPSO after the falling edge of WRCK. The WPSO CTL code follows.

```
<4.492>   Environment EX_wrapper {
<4.493>     CTLMode {
<4.494>       Internal {
<4.495>         WPSO[0..3] {
<4.496>           DataType TestData ScanDataOut {
<4.497>             ScanDataType Boundary;
<4.498>           }
<4.499>           Wrapper User WPP PinID WPO;
```

```
                    StrobeRequirements {
                        TimingSensitive {
                            Reference WRCK {
                                SelfEdge LeadingTrailing;
                                ReferenceEdge Trailing;
                                EarliestTimeValid <tsov>;
                            } // Reference block
                        } // TimingSensitive block
                    } // StrobeRequirements block
                } // WPSO[0..3] block
            } // Internal block
        } // CTLMode block
    } // Environment block
```

t_{sov} represents the valid output time corresponding to Figure 24

4.3.11.2 Declaring WPO for the EX wrapper MBIST mode

Three WPO terminals named MBISTFAIL, MBISTDONE, and MBISTL-OGOUT will be used during MBIST. The timing corresponding to these terminals is based on the functional (CLK) clock given that MBIST operates functional storage elements inside the core. The "WPO" timing diagram is shown in Figure 25.

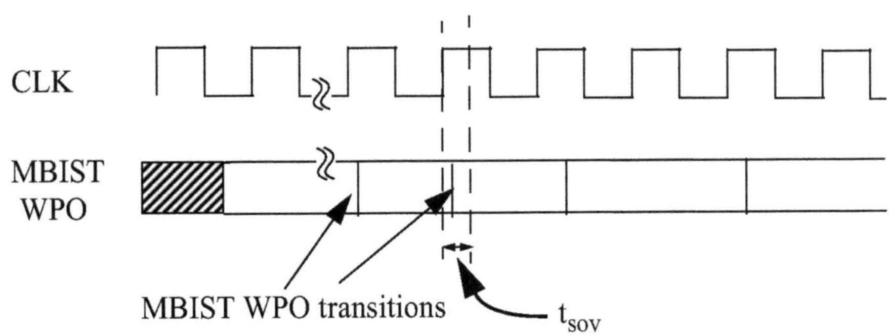

Figure 25 Timing of the EX wrapper MBIST WPO terminals

t_{sov} represents the time is takes for data to be valid after the rising edge of CLK. The corresponding CTL code follows.

```
Environment EX_wrapper {
    CTLMode {
        Internal {
```

```
            MBISTFAIL {
                DataType TestData ScanDataOut {
                    ScanDataType Boundary;
                }
                Wrapper User WPP PinID WPO;
                StrobeRequirements {
                    TimingSensitive {
                        Reference CLK {
                            SelfEdge LeadingTrailing;
                            ReferenceEdge Leading;
                            EarliestTimeValid <t_sov>;
                        } // Reference block
                    } // TimingSensitive block
                } // StrobeRequirements block
            } // MBISTFAIL block
            MBISTDONE {
                DataType TestData ScanDataOut {
                    ScanDataType Boundary;
                }
                Wrapper User WPP PinID WPO;
                StrobeRequirements {
                    TimingSensitive {
                        Reference CLK {
                            SelfEdge LeadingTrailing;
                            ReferenceEdge Leading;
                            EarliestTimeValid <t_sov>;
                        } // Reference block
                    } // TimingSensitive block
                } // StrobeRequirements block
            } // MBISTDONE block
            MBISTLOGOUT {
                DataType TestData ScanDataOut {
                    ScanDataType Boundary;
                }
                Wrapper User WPP PinID WPO;
                StrobeRequirements {
                    TimingSensitive {
                        Reference CLK {
                            SelfEdge LeadingTrailing;
                            ReferenceEdge Leading;
                            EarliestTimeValid <t_sov>;
                        } // Reference block
                    } // TimingSensitive block
                } // StrobeRequirements block
            } // MBISTLOGOUT block
        } // Internal block
    } // CTLMode block
} // Environment block
```

t_{sov} represents the valid output time corresponding to Figure 25

4.4 Combined CTL Code for EX Wrapper Terminals

As we have completed CTL code description for individual EX wrapper interface terminals, we can look at the combined 1500 compliant CTL code describing this interface. The CTL code follows.

```
STIL 1.0 {
Design 2005;
CTL 2005;
}
Signals {
    WRCK In;
    WRSTN In;
    SelectWIR In;
    ShiftWR In;
    CaptureWR In;
    UpdateWR In;
    WSI In;
    WPSI[0..3] In;
    WPSE In;
    WSO Out;
    WPSO[0..3] Out;
    CLK In;
    RESET In;
    ADDR[0..5] In;
    DIN[0..7] In;
    DOUT[0..7] Out;
    READY In;
    BC Out;
    ACK Out;
    RX Out;
    TX Out;
    MBISTDLOG In;
    MBISTRUN In;
    MBISTDLOGOUT Out;
    MBISTDONE Out;
    MBISTFAIL Out;
}
```

```
<4.596>  SignalGroups {
<4.597>     all_functional='CLK + RESET + ADDR[0..5] + DIN[0..7] +
               DOUT[0..7] + READY + BC + ACK + RX + TX';
<4.598>     wpp_mbist='MBISTDLOG + MBISTRUN +
            MBISTDLOGOUT + MBISTDONE + MBISTFAIL + CLK';
<4.599>     wpp_scan='WPSI[0..3] + WPSO[0..3] + WPSE + CLK';
<4.600>     wpp_wrck='WRCK';
<4.601>     wsp_wrck='WRCK';
<4.602>  }
<4.603>  Environment EX_wrapper {
<4.604>     CTLMode {
<4.605>        Internal {
<4.606>           wsp_wrck {
<4.607>              DataType TestControl ScanMasterClock;
<4.608>              Wrapper IEEE1500 PinID WRCK;
<4.609>              DriveRequirements {
<4.610>                 TimingSensitive {
<4.611>                 Pulse High Min <t_ckwh>;
<4.612>                 Pulse Low Min <t_ckwl>;
<4.613>                 Period Min <t_ckwh> + <t_ckwl>;
<4.614>                 } // TimingSensitive block
<4.615>              } // DriveRequirements block
<4.616>           } // wsp_wrck block
<4.617>           wpp_wrck {
<4.618>              DataType TestControl ScanMasterClock;
<4.619>              Wrapper User WPP PinID WPC;
<4.620>              DriveRequirements {
<4.621>                 TimingSensitive {
<4.622>                 Pulse High Min <t_ckwh>;
<4.623>                 Pulse Low Min <t_ckwl>;
<4.624>                 Period Min <t_ckwh + t_ckwl>;
<4.625>                 } // TimingSensitive block
<4.626>              } // DriveRequirements block
<4.627>           } // wpp_wrck block
<4.628>           WRSTN {
<4.629>              DataType TestControl TestWrapperControl {
<4.630>                 ActiveState ForceDown;
<4.631>              }
<4.632>              Wrapper IEEE1500 PinID WRSTN;
```

```
        DriveRequirements {
            TimingSensitive {
                Pulse Low Min <t_rstl>;
                Reference WRCK {
                    SelfEdge Trailing;
                    ReferenceEdge Leading;
                    Setup <t_rstsu>;
                } // Reference block
            } // TimingSensitive block
        } // DriveRequirements block
    } // WRSTN block
    SelectWIR {
        DataType TestControl TestWrapperControl {
            ActiveState ForceUp;
        }
        Wrapper IEEE1500 PinID SelectWIR;
        DriveRequirements {
            TimingSensitive {
                Reference WRCK {
                    SelfEdge LeadingTrailing;
                    ReferenceEdge Leading;
                    Setup <t_swsu>;
                    Hold <t_swhd>;
                } // Reference block
            } // TimingSensitive block
        } // DriveRequirements block
    } // SelectWIR block
    ShiftWR {
        DataType TestControl ScanEnable;
        DataType TestControl TestWrapperControl {
            ActiveState ForceUp;
        }
        Wrapper IEEE1500 PinID ShiftWR;
        DriveRequirements {
            TimingSensitive {
                Reference WRCK {
                    SelfEdge LeadingTrailing;
                    ReferenceEdge Leading;
                    Setup <t_ctlsu>;
                    Hold <t_ctlhd>;
                } // Reference block
            } // TimingSensitive block
        } // DriveRequirements block
    } // ShiftWR block
    CaptureWR {
        DataType TestControl TestWrapperControl {
            ActiveState ForceUp;
        }
        Wrapper IEEE1500 PinID CaptureWR;
```

```
    DriveRequirements {
        TimingSensitive {
            Reference WRCK {
                SelfEdge LeadingTrailing;
                ReferenceEdge Leading;
                Setup <t_ctlsu>;
                Hold <t_ctlhd>;
            } // Reference block
        } // TimingSensitive block
    } // DriveRequirements block
} // CaptureWR block
UpdateWR {
    DataType TestControl TestWrapperControl {
        ActiveState ForceUp;
    }
    Wrapper IEEE1500 PinID UpdateWR;
    DriveRequirements {
        TimingSensitive {
            Reference WRCK {
                SelfEdge LeadingTrailing;
                ReferenceEdge Trailing;
                Setup <t_updsu>;
                Hold <t_updhd>;
            } // Reference block
        } // TimingSensitive block
    } // DriveRequirements block
} // UpdateWR block
CLK {
    DataType TestControl ScanMasterClock;
    Wrapper User WPP PinID WPC;
    DriveRequirements {
        TimingSensitive {
            Reference WRCK {
                SelfEdge Leading;
                ReferenceEdge Leading;
                Setup <t_ctlsu>;
                Hold <t_ctlhd>;
            } // Reference block
        } // TimingSensitive block
    } // DriveRequirements block
} // CLK block
WSI {
    DataType TestData ScanDataIn {
        ScanDataType Boundary;
    }
    Wrapper IEEE1500 PinID WSI;
```

```
        StrobeRequirements {
            TimingSensitive {
                Reference WRCK {
                SelfEdge LeadingTrailing;
                ReferenceEdge Trailing;
                EarliestTimeValid <tsov>;
                } // Reference block
            } // TimingSensitive block
        } ) // StrobeRequirements block
    } // WSI block
    WSO {
        DataType TestData ScanDataOut {
            ScanDataType Boundary;
        }
        Wrapper IEEE1500 PinID WSO;
        StrobeRequirements {
            TimingSensitive {
                Reference WRCK {
                    SelfEdge LeadingTrailing;
                    ReferenceEdge Trailing;
                    EarliestTimeValid <tsov>;
                } // Reference block
            } // TimingSensitive block
        } // StrobeRequirements block
    } // WSO block
    WPSI [0..3] {
        DataType Testdata ScanDataIn {
            ScanDataType Boundary;
        }
        Wrapper User WPP PinID WPI;
        DriveRequirements {
            TimingSensitive {
                Reference WRCK {
                    SelfEdge LeadingTrailing;
                    ReferenceEdge Leading;
                    Setup <tsisu>;
                    Hold <tsihd>;
                } // Reference block
            } // TimingSensitive block
        } // DriveRequirements block
    } // WPSI [0..3] block
    WPSO[0..3] {
        DataType TestData ScanDataOut {
            ScanDataType Boundary;
        }
        Wrapper User WPP PinID WPO;
```

```
            StrobeRequirements {
                TimingSensitive {
                    Reference WRCK {
                        SelfEdge LeadingTrailing;
                        ReferenceEdge Trailing;
                        EarliestTimeValid <t_sov>;
                    } // Reference block
                } // TimingSensitive block
            } // StrobeRequirements block
        } // WPSO[0..3] block
        MBISTFAIL {
            DataType TestData ScanDataOut {
                ScanDataType Boundary;
            }
            Wrapper User WPP PinID WPO;
            StrobeRequirements {
                TimingSensitive {
                    Reference CLK {
                        SelfEdge LeadingTrailing;
                        ReferenceEdge Leading;
                        EarliestTimeValid <t_sov>;
                    } // Reference block
                } // TimingSensitive block
            } // StrobeRequirements block
        } // MBISTFAIL block
        MBISTDONE {
            DataType TestData ScanDataOut {
                ScanDataType Boundary;
            }
            Wrapper User WPP PinID WPO;
            StrobeRequirements {
                TimingSensitive {
                    Reference CLK {
                        SelfEdge LeadingTrailing;
                        ReferenceEdge Leading;
                        EarliestTimeValid <t_sov>;
                    } // Reference block
                } // TimingSensitive block
            } // StrobeRequirements block
        } // MBISTDONE block
        MBISTLOGOUT {
            DataType TestData ScanDataOut {
                ScanDataType Boundary;
            }
            Wrapper User WPP PinID WPO;
```

```
            StrobeRequirements {
                TimingSensitive {
                    Reference CLK {
                        SelfEdge LeadingTrailing;
                        ReferenceEdge Leading;
                        EarliestTimeValid <t_sov>;
                    } // Reference block
                } // TimingSensitive block
            } // StrobeRequirements block
        } // MBISTLOGOUT block
        MBISTRUN {
            DataType TestControl TestWrapperControl;
            Wrapper User WPP PinID WPC;
            DriveRequirements {
                TimingSensitive {
                    Reference CLK {
                        SelfEdge LeadingTrailing;
                        ReferenceEdge Leading;
                        Setup <t_ctlsu>;
                        Hold <t_ctlhd>;
                    } // Reference block
                } // TimingSensitive block
            } // DriveRequirements block
        } // MBISTRUN block
        MBISTLOG {
            DataType TestControl TestWrapperControl;
            Wrapper User WPP PinID WPC;
            DriveRequirements {
                TimingSensitive {
                    Reference CLK {
                        SelfEdge LeadingTrailing;
                        ReferenceEdge Leading;
                        Setup <t_ctlsu>;
                        Hold <t_ctlhd>;
                    } // Reference block
                } // TimingSensitive block
            } // DriveRequirements block
        } // MBISTLOG block
        WPSE {
            DataType TestControl TestWrapperControl;
            DataType TestControl ScanEnable {
                ActiveState ForceValid;
            }
            Wrapper User WPP PinID WPC;
```

```
                DriveRequirements {
                    TimingSensitive {
                        Reference WRCK {
                            SelfEdge LeadingTrailing;
                            ReferenceEdge Leading;
                            Setup <t_ctlsu>;
                            Hold <t_ctlhd>;
                        } // Reference block
                    } // TimingSensitive block
                } // DriveRequirements block
            } // WPSE block
        } // Internal block
    } // CTLMode block
} // Environment block
```

The above CTL code will be incrementally built upon as additional components of the 1500 architecture are covered in the following chapters.

4.5 Plug-and-play of the 1500 Architecture

From a core test perspective plug-and-play refers to an ability that allows multiple 1500 wrappers to operate when connected serially, even though these wrappers may have originated from different core providers. The primary requirement for plug-and-play in this context is that there must be data bandwidth matching between the various cores connected together. This prevents plug-and-play ability in tests that make use of the WPP since the WPP is user-defined by nature. Therefore, plug-and-play is limited to serial configurations of the wrapper. Furthermore, user-defined operations of the wrapper cannot be expected to be plug-and-play although they may be if these operations are carefully defined. The standard guarantees plug-and-play through a serial external instruction (WS_EXTEST). WS_EXTEST is defined in Section *5.1.1.2* (see page 80).

Assuming that 1500 Plug-and-Play requirements are met for each individual core, timing requirements from WSO to WSI of any two wrappers serially connected must be guaranteed at the SOC level. As a result, when wrappers functioning at different frequencies are serially connected, the frequency of operation of the slowest wrapper in this connection will define the frequency of operation of the chain of wrappers. Falling edge changing of the WSO terminal, and rising edge sampling of the WSI terminal were introduced to help meet WSO to WSI timing requirements for Plug-and-Play purposes.

The use of functional storage elements in the WBR creates a hurdle for Plug-and-Play since this typically indicates that WRCK will not directly operate these storage elements. This takes away from the ability to expect data to propagate in the WBR cells of all wrappers connected serially, unless the auxiliary clock operating the functional storage element is synchronized to WRCK as described in Section *4.1.1.9* (see page 35).

If any other external instruction besides WS_EXTEST is utilized, the responsibility of plug-and-play resides with the SOC designer.

Chapter 5
Instruction Types

The instruction set defined by 1500 comprises instructions that were anticipated would be the most utilized instructions or would help guide the 1500 standard users to create their own instructions. The instruction set defined in the standard contains both mandatory and optional instructions. The mandatory instructions cover the absolute minimum set of instructions that each 1500 compliant wrapper must provide. These instructions allow the core and wrapper to execute a functional mode, an internal test mode and an external test mode. The optional instructions defined provide mechanisms for additional abilities by changing the configuration and capability of the WDRs. In addition to the instructions defined by the 1500 standard, there is an undefined class of instructions called user-defined instructions. This chapter will discuss all three types of instructions, and show how and why they might be used. Each instruction must have an opcode that enables it. While the opcode is mandatory, the value of the opcode can be defined for each instruction by the designer. An opcode is a unique instruction or test mode identifier that gets shifted into the WIR in preparation of the core and wrapper going into the corresponding test mode.

5.1 Overview of the IEEE 1500 Standard Instructions

The 1500-defined instruction set is composed of:

- serial instructions - where the data uses only a serial path (WSI to WSO) and the wrapper serial controls (WSC)
- parallel instructions - where the data may use a Test Access Mechanism (TAM) of 0 to n bits width in conjunction with user-defined controls
- hybrid instructions - where a combination of the terminals for serial and parallel instructions can be utilized.

There is a naming convention for the instructions that provides some insight into what the instruction does. The instruction name is broken up into the following three sections separated by an underscore (_):

1. The first section contains two letters. The first letter is always W. This W was initially added to differentiate 1500 instructions from 1149.1 instructions as they have many similarities. The second letter is an S, a P or an H. This distinguishes whether the instruction is a serial, a parallel or a hybrid instruction.

- A serial instruction utilizes solely the WSP during the test.

- A parallel instruction utilize solely user-defined test ports (no WSP signals), referred to as the WPP, with the exception of WP_EXTEST instruction. See Section *5.1.1.2* (see page 80).

- A hybrid instruction utilizes a combination of the WSP and the WPP to access the WDRs and run its test.

  With the exception of WP_EXTEST, a parallel instruction cannot use any of the WSP terminals during the test with the exception of the WRCK and auxiliary clocks. WP_EXTEST may require a more restrictive protocol as it will be communicating with the WDRs of other cores or UDL. For this reason, the wrapper serial control (WSC) signals can be utilized by WP_EXTEST.

  For flexibility, the hybrid instruction capability was included. For example, if a test mode instruction requires the use of ShiftWR to shift data through any scan chain and it also uses scan ports that are different from WSI and WSO, then this is a hybrid instruction. This type of instruction was added to give maximum flexibility to the 1500 standard user, while remaining within the frame of a structured instruction.

  Of course, the WIR and WBY can only be accessed by the WSP. It should be noted that some of the instructions described in the standard contain an x in the second letter position (i.e Wx). This indicates that the instruction can be a serial, parallel or hybrid instruction. During actual use of the instruction, the x would be replaced by an S, a P or an H as is appropriate. There would never be an implemented instruction that begins with Wx.

2. The second section of the instruction name contains the main function of the instruction. For instance, INTEST would indicate that an inter-

nal test is in operation. EXTEST would indicate an external test mode. BYPASS would indicate that the WBR or any other WDR is being bypassed.

3. The third section of the instruction name is optional and is used to provide a little extra description or differentiation from any other instruction. This third section is utilized in the WS_INTEST_SCAN and WS_INTEST_RING instructions. These instructions are explained in Section *5.1.2.7* (see page 88) and Section *5.1.2.8* (see page 88). As can be seen, they are both serial internal test instructions. However, WS_INTEST_SCAN incorporates WBR scan chains as well as core internal scan chains in its single scan chain. Whereas only the WBR scan chains are concatenated together to create a single scan chain during the execution of the WS_INTEST_RING instruction.

5.1.1 Mandatory Instructions

There are three mandatory instructions in the 1500 standard. These three mandatory instructions cover the basic needs of a core that will be placed in an SOC. A functional mode, an external test mode and an internal test mode. A core can have as many instructions as required to have a complete test, but must always have the three mandatory instructions included in that set. Table 5 lists the three mandatory instructions.

Table 5 IEEE 1500 Instruction List

Instruction	Mandate	Description
WS_BYPASS	Required	Allows mission (functional) mode to occur and puts the wrapper into bypass mode.
WS_EXTEST	Required	Allows external test using a single chain configuration in the WBR.
Wx_INTEST	Required	Allows freedom to drive the internal testing needed for a core. Encompasses all of the INTEST instructions described in the 1500 standard.

5.1.1.1 WS_BYPASS

The first mandatory instruction that will be discussed is WS_BYPASS. WS_BYPASS configures the WBR into transparent mode. In other words, all of the functional signals pass right through the WBR as if it were not there. The WBY is connected between WSI and WSO during this instruction. In addition, the wrapped core is put into functional mode. WS_BYPASS has two main purposes and would be used in two completely different scenarios. The first utilization is to put the core into normal or functional mode. In this case, the WBY path from WSI to WSO is typically ignored. The second use is to bypass the WBR to shorten the WDR length during a test mode as described in Chapter 7. Functional signals coming through the WBR are probably ignored, though there is no reason that they cannot be utilized in some way. This is up to the SOC developer. While this serial instruction is being utilized, only the WSP may be used to input and output data and control to the wrapper. Note that while the core is executing in functional mode, the WBY path can also be used.

WS_BYPASS is the default instruction. During reset of the core, this instruction is forced into the WIR utilizing the WRSTN. By the time the WRSTN is disabled, all signals being driven from the WIR to the core, WBR and WBY, must put the core and WBR into functional mode and put the WBY in the WSI to WSO path as described in Chapter 7. Let's refer to the unwrapped example core EX in Chapter 3. In order for EX to be in functional mode, SCANMODE and MBISTMODE must be disabled (set to 0). All of the other test signals are gated with SCANMODE and MBISTMODE, so these can be set to whatever is convenient. When WRSTN is enabled (set to 0), SCANMODE and MBISTMODE are driven to 0 and remain in that state when WRSTN is disabled. The WBR must be in transparent mode during this instruction. The transparent mode is also referred to as the Wrapper Disabled State as described in Section *8.2.2* (see page 191).

5.1.1.2 WS_EXTEST

The WS_EXTEST instruction provides the simplest interface to test logic external to the core. Since all serial instructions can only utilize the WSP and since all cores must implement the WS_EXTEST instruction mode, this provides the simplest interface to all cores to test logic adjacent to these cores. It also creates the most fail-safe path to implement Plug-and-Play between the cores. Plug-and-Play characteristics are described in Section *4.5* (see page 75). Multiple wrappers may need to be exercised to test logic external to the core or cores. It is important that they can all run synchronously in order to be used in conjunction with each other.

The user-defined logic (UDL) shown in Figure 26 is connected to both Core1 and Core2. This means that both cores' wrappers must be utilized to test this logic. As can be seen, the WSO of Core1 is connected to the WSI of Core2. This is expected to be a common scenario as it allows all of the WIRs to be concatenated as well as the WBRs when needed. The WSC signals go to both cores. This is also an expected scenario on an SOC instantiating 1500 cores as it allows all cores to operate together. Every core will load the instruction into the WIR at the same time. Every core can update the WIR to transfer that instruction to the core and WDRs, then all of the WBRs can shift, apply and capture together to test logic external to the core.

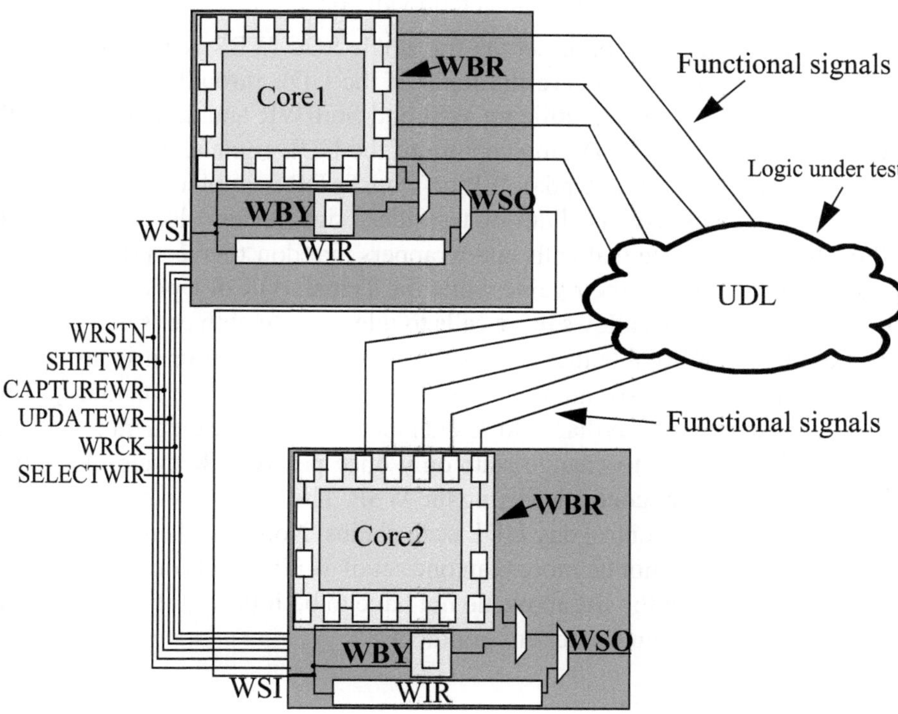

Figure 26 WS_EXTEST Example

WS_EXTEST configures the WBR into Outward Facing mode as described in Section *6.5.1* (see page 152).

While both cores in Figure 26 are in WS_EXTEST instruction mode, the following setup is required per core:

1. The WBR is in Outward Facing mode
2. The WBR is connected between WSI and WSO
3. All scan chains in the WBR are concatenated together - a single wrapper chain.

and the following setup is recommended in the SOC:

4. WSO of a preceding core is connected to the WSI of the subsequent core.
5. The WSC signals are connected to all cores.

Once this configuration is set up, the single SOC WBR scan chain can be utilized to control and observe the logic of the UDL through the WSP. The WBR scan chain is loaded utilizing WSI and ShiftWR and unloaded through WSO utilizing ShiftWR. During capture of the UDL signals, the CaptureWR signal is enabled. If either UpdateWR or TransferDR is implemented in any of the wrappers being exercised during test, these signals may also need to go to all of the WBRs. Note that cells and wrappers that don't have update/transfer need to be able to hold their value while the TransferDR or UpdateWR signal is being asserted. This allows these cells to interoperate with cells that do have update/transfer. WRCK either clocks the WBR scan chain or it is an enable for a functional clock (or auxiliary clock as it is called in the standard). The WRCK must gate the auxiliary clock such that each storage element inside the WBR cell is allowed to change state only once per WRCK cycle. All functions of the WBR are controlled using the WSP. The UDL may require signals beyond the WSP to control any UDL scan chains. Note that there are no rules stating that there cannot be more than one set of WSP signals on an SOC. The recommendations in the list above do not have to be followed if another setup better suits the SOC and tester environment.

5.1.1.3 Wx_INTEST

The last mandatory instruction is Wx_INTEST. Actually, any internal test (INTEST) instruction can be employed to fill this requirement. What is mandatory is that there be at least one INTEST instruction that can test the core.

Since the INTEST instruction can be totally defined by the core and wrapper designer, it can be a serial, parallel or hybrid instruction. The WBR must be configured in Inward Facing mode as described in Section *6.5.1* (see page 152). This instruction allows the core internal test to be defined by the core provider with virtually no restrictions on the type of test implemented. A few examples of types of test include scan, BIST and functional. This flexibility is needed for test methodologies today and in the future.

The pre-defined WS_INTEST_SCAN and WS_INTEST_RING instructions are included in the Wx_INTEST description. If either one of these instructions are implemented, the Wx_INTEST requirement is met. If this is the case, one might wonder why the WS_INTEST_SCAN and WS_INTEST_RING were defined at all in the 1500 standard. The reason behind this is that, if standard users did implement these type of tests, there should be a more defined structure to the instructions that accompany them. This may make test generation more amenable to automation. Serial instructions are more rigorous in their rules to ensure that they allow Plug-and-Play.

5.1.2 Optional Instructions

The 1500 standard defines a set of optional instructions to facilitate some commonly needed operations for testing cores and SOCs. They were specifically created to help the 1500 standard user define these types of actions. It also creates a standard way for these instructions to be carried out. This helps the entire test community in a couple of ways. The first way is that if a core is received with any of the optional instructions, the implementation of the instruction is understood. This makes integration easier. The second way that it helps is that it makes it easier to create automation for the hardware and software needs for these instructions.

Table 6 shows a list of the 1500 optional instructions

Table 6 IEEE 1500 Optional Instructions

Instruction	Description
WP_EXTEST	External test mode instruction that utilizes the WPP and possibly the WSC
WS_SAFE	Statically drives predetermined (hardwired) safe values from wrapper functional outputs and configures the WBY to be between WSI and WSO.
WS_CLAMP	Places the WBR in OF mode and configures the WBY to be between WSI and WSO. Presumed to be preceded by WS_PRELOAD.
WP_PRELOAD	Loads data into the dedicated shift path of the WBR through the WPP.
WS_PRELOAD	Loads data into the dedicated shift path of the WBR through the WSP. Mandatory if there is a silent shift path.
WS_INTEST_RING	Allows internal testing using a single chain configuration in the WBR.
WS_INTEST_SCAN	Allows internal testing by concatenation of the wrapper chain with a single internal chain.

The term silent shift path is used in Table 6. A silent shift path allows data to be shifted into and out of it without disturbing the core functional outputs or the core functional inputs in every wrapper mode. As stated, the optional instructions are defined to help create a standard method of utilizing the wrapper in ways that were thought to be most commonly needed. There will be a brief description of each instruction and the reason it might be included in the core's repertoire in the following sections.

5.1.2.1 WP_EXTEST

There is a mandatory serial external test instruction (WS_EXTEST). However, many SOC designers may desire to shorten their test time and lessen the tester memory depth needs by shortening the length of the scan chains. In this case, the WS_EXTEST instruction may not meet their needs as it requires that the WBR be composed of a single scan chain, which may be quite long. The WP_EXTEST instruction was included in order to have a standard way of accessing multiple scan chains in the WBR. WP_EXTEST is the only parallel instruction that can utilize the WSC signals as shown in Figure 27 (page 85). Since external test mode requires the most handshaking (Plug-and-Play) between cores, it was decided that this was an exception that should be made and even encouraged. The wrapper parallel scan input and output (WPI and WPO) must be utilized during the WP_EXTEST instruction. WSI and WSO are connected to the WBY during execution of this instruction. Note that the WBR cells depicted in Figure 27 do not show the mandatory hold behavior required by the 1500 standard. The hold capability can be obtained in multiple ways e.g. gating off the clock. This figure represents the scan path of a WBR chain in parallel mode. Although a single WBR cell is shown, the scan path will typically contain multiple scan cells. The input to the scan chain is WPI while the output is WPO.

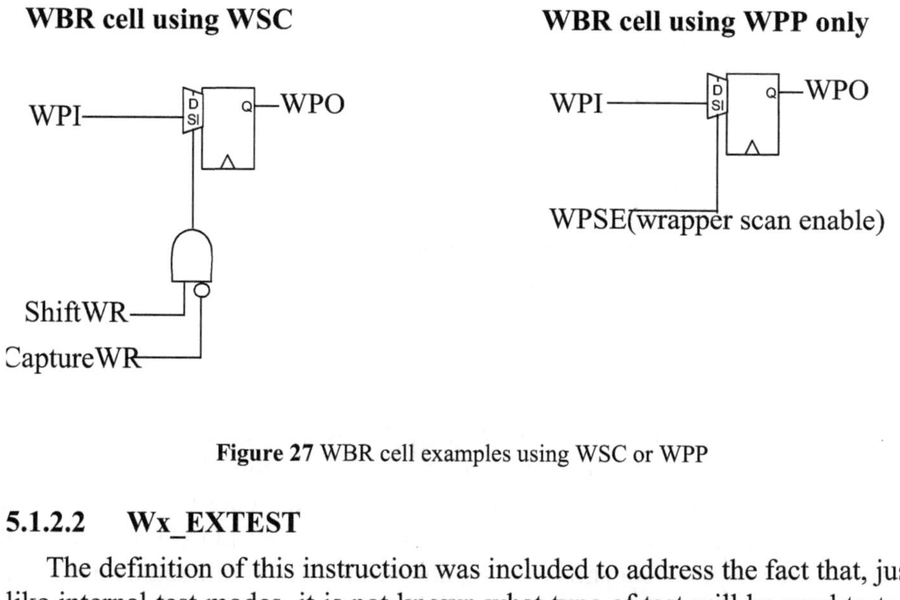

Figure 27 WBR cell examples using WSC or WPP

5.1.2.2 Wx_EXTEST

The definition of this instruction was included to address the fact that, just like internal test modes, it is not known what type of test will be used to test the logic external to the core. The Wx_EXTEST instruction encompasses WS_EXTEST and WP_EXTEST, but also includes any other type of external test mode that follows the 1500 rules. This allows virtually any type of external test to be employed, but still has some standardization through the rules defined.

5.1.2.3 WS_SAFE

The purpose of WS_SAFE is to put all of the core outputs into a "safe" state. In this case safe equates to a static signal that will not damage the logic external to the core. This instruction was added to allow a way to quickly put the core outputs into a known state, without having to shift any values into the WBR. For this instruction, the outputs would be gated to immediately produce these values once the WS_SAFE instruction is engaged. Figure 28 shows an example of a safe gate that outputs a 0 when enabled. Note that every single output must have a "safe gate" on it in order to be compliant with this instruction.

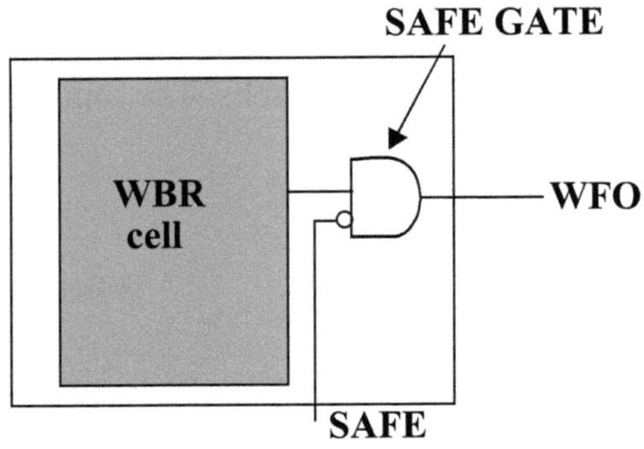

Figure 28 Example of a WBR cell with a safe gate

The input signal gating is optional as is putting the core into reset. This is a serial instruction, so only the WSP can be utilized during the execution of this instruction.

5.1.2.4 WS_PRELOAD

WS_PRELOAD is a serial instruction that was created to allow a WBR shift stage to be filled with data before a test is begun. This instruction is required if there is a silent shift path. The preloaded data is updated to the functional output by the instruction following WS_PRELOAD. Every WBR cell must have the silent shift path capability in order to utilize this particular instruction. In fact, if every WBR cell does have the silent shift path capability, this instruction is mandated.

The WC_SD1_COI_UD WBR cell, shown in Figure 29, is an example of a WBR cell that incorporates the silent shift path capability. This cell has a dedicated shift register and an update register, so will not disturb the core state or external logic state during shift. A single dedicated shift storage element is utilized while the WBR is in a test configuration and this storage element never participates in the functional operation of the core. An update storage element protects the CFO of the WBR cell from being changed during shift of the dedicated shift register.

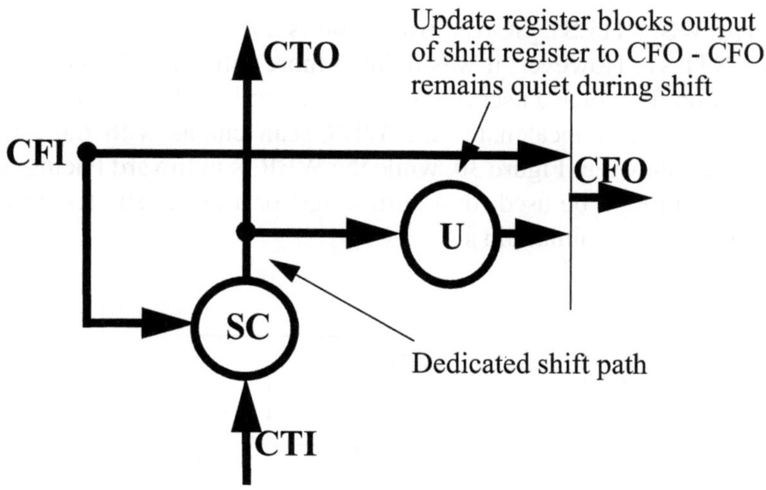

Figure 29 WC_SD1_COI_UD WBR Cell

5.1.2.5 WP_PRELOAD

The WP_PRELOAD serves the same purpose as the WS_PRELOAD except that it utilizes the WPP. The reason for this instruction is to allow access to multiple scan chains in the WBR simultaneously, if desired, rather than the single scan chain required for WS_PRELOAD. This can make the preload much shorter. In addition, if multiple scan chains are being accessed during WP_EXTEST, the same scan chains can be accessed during WP_PRELOAD. WP_PRELOAD may directly precede an instruction such as WP_EXTEST if known values are needed in the WBR at the start of the WP_EXTEST test.

5.1.2.6 WS_CLAMP

The WS_CLAMP instruction might be used after WS_PRELOAD or WP_PRELOAD to clamp the data to the outputs of a core. The Wx_PRELOAD instruction and the WS_CLAMP instruction can be used in conjunction to do the exact same thing as WS_SAFE, except this allows any value to be shifted into the WBR cell and clamped to the output. This is much more flexible than the WS_SAFE. However, the WS_SAFE instruction allows for a much less complicated way to get the "safe" values on the outputs of a core - provided only one set of values is needed on the outputs during test and the logic of the gate in the functional path is not an issue.

5.1.2.7 WS_INTEST_SCAN

The WS_INTEST_SCAN instruction is encompassed in the definition of Wx_INTEST. However, it was defined more fully and stringently as a serial instruction for which a user of the 1500 standard might have a need. This is an instruction that concatenates the WBR scan chains with the internal scan chains, as shown in Figure 30, while the WBR is in Inward Facing mode. This instruction might be used for a burn-in test or a diagnostic test or an internal test utilizing a minimal pin set.

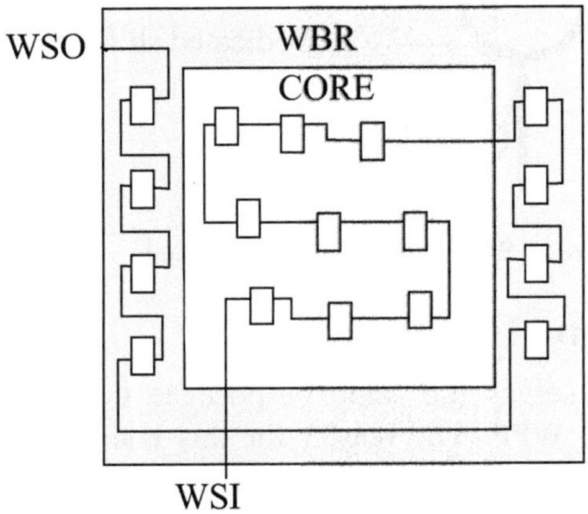

Figure 30 WBR chain(s) concatenated with internal chain(s)

Since this is a serial instruction, only the WSP can be used to execute this test.

5.1.2.8 WS_INTEST_RING

The WS_INTEST_RING is encompassed in the definition of Wx_INTEST. It follows all of the same rules as WS_INTEST_SCAN, except that only the WBR is concatenated into a single chain. The internal scan chains are not incorporated into this test. One possible use of this mode is if the core has no scan chains and functional patterns are needed to test the core. The WBR can be set up in a way that can deliver those functional patterns cycle by cycle. The WBR cells can be more than one register deep allowing more than one bit of data to be read into the core after shift is complete.

5.1.3 User-Defined Instructions

The mandatory and optional instructions defined in the standard may not be all that is needed by the 1500 standard user. The user may need a completely different instruction or a variation of a mandatory or optional instruction. In this case, a user-defined instruction can be created.

For instance, the WS_SAFE instructions requires that every output be gated in order to be compliant. A possible scenario may be that only one signal must have a "safe" output on a core. The wrapper designer may not desire or be able to put a gate on every output of the core. In this case, a user-defined instruction could be created that controlled that one output's safe gate.

It is recommended that the user-defined instructions follow the same naming convention as the standard defined instructions. The user should follow the rules stated for the first and second section of the instruction and can add the third section if needed. No defined 1500 mandatory or optional instruction names can be used for the user-defined instructions. However, derivatives of the mandatory and optional instruction names may be used. For instance, for the user-defined instruction described in this section, an appropriate name may be WS_SAFE_SINGLE.

Another example of a possible user-defined instruction could be a derivative of WS_PRELOAD or WP_PRELOAD. The PRELOAD instructions mandate that the silent shift path include all of the wrapper cells. If there is one that does not meet this requirement, the WS_PRELOAD or WP_PRELOAD instructions are not mandated and cannot be used. However, if the standard user has a silent shift capability on some the WBR cells included in the wrapper, a user preload instruction can be created. If the instruction is a serial instruction, it may have a name such as WS_PRELOAD_PARTIAL.

5.2 EX Wrapped Core Instructions

In order to build the EX wrapper, instructions must be defined. It is already known that this instruction set must include WS_EXTEST, WS_BYPASS and at least one Wx_INTEST instruction. The EX core has three different internal test modes. A scan mode with a single scan chain, a scan mode with multiple scan chains and a memory BIST mode. Three instructions will be created to implement these three modes - WS_INTEST, WP_INTEST, and WP_INTEST_MBIST. The first two instructions are standard defined instructions. The third instruction, WP_INTEST_MBIST, is a user-defined instruction.

WS_INTEST follows all of the rules of WS_INTEST_SCAN except Rule 7.13.1(a) in the 1500 standard, which states that a single internal scan chain is concatenated to a single WBR chain. For the EX core, this is set up as shown in Figure 102 (page 212). This figure shows a single serial scan chain, with internal chains inter-mixed with the WBR chains, and also shows that the EX core has configurable wrapper chains. With this implementation there can be a single scan chain or there can be four separate scan chains in the WBR. The WP_EXTEST instruction is added to accommodate the four chain WBR configuration - another standard defined instruction.

The EX core uses a WS_SAFE user-defined instruction to prevent contention on the logic being controlled by BC. This instruction adds an extra gate to the functional path as shown in Figure 28 (page 86), so timing must be taken into consideration. In the case of the EX core, timing is not critical on the BC signal and a user-defined WS_SAFE instruction is utilized. The new user-defined instruction will be called WS_SAFE_SINGLE.

Table 7 shows the instructions that will be used on the EX core. The name of the instruction is listed in the first column. The second column displays whether the instruction is mandatory or not. The third column indicates whether it is a standard defined instruction or a user-defined instruction and the last column contains a brief description of the configuration of the wrapper during that instruction.

There are two more user-defined instructions that the EX core utilizes; WP_INTEST_SEQ and WP_EXTEST_SEQ. These two instructions were created so that sequential data could be input to the logic being tested, from the wrapper. WP_INTEST_SEQ configures the WBR into Inward Facing mode and is basically the same as the EX core WP_INTEST instruction, with one exception. During the WP_INTEST_SEQ instruction the scan enable to all of the input WBR cells is held in shift mode, while the scan enable to all of the output WBR cells is allowed to capture data as is normal. During internal test mode, the input WBR cells are allowed to function normally, they can only hold their previous data. If the WBR cells remain in shift mode, dynamic data can be applied to the core logic. This is essential for sequential testing. For instance, delay testing requires data pairs (1 to 0 or 0 to 1) to be applied to logic.

The logic connected to the functional inputs can be delay tested with this methodology, without using WBR cells with multiple storage elements. Another type of sequential test that this methodology may enable is testing through memories. Some memories are not wrapped inside of a core. This means that in order to test the shadow logic around the memory, the tool must test through the memory. This could require several sequential bits of data. If the memory is sequentially close to an input port, it may require this sequential data to come from the WBR. The methodology described here can accommodate this requirement.

WP_EXTEST_SEQ works in the same way as WP_INTEST_SEQ, except it is applied to logic external to the core during external test mode. During this instruction all of the output WBR cells are held in shift mode and the input WBR cells are allowed to capture. 1500 rule 12.3.1.1(a) states that the Shift event shall not occur simultaneously with either Transfer, Capture, or Update events. This means that during the WBR Shift event, none of the other stated events can occur. This is the case during the WBR Shift event of the WP_INTEST_SEQ instruction. However, during capture, no such rule exists. During capture the WP_INTEST_SEQ instruction allows some of the WBR cells to be in shift mode and others to be in capture mode. In other words, the specificity of this instruction comes from the fact that it overlaps the capture event and the apply event that provides stimulus data to the core inputs. The overlapping of events is allowed by the standard.

An alternative way of supporting sequential test with the 1500 architecture is discussed in subclause 12.8 of the standard. This method makes use of WBR cells with multiple storage elements.

One final instruction that is utilized on the EX core is the WP_BYPASS instruction. While WP_BYPASS is enabled, each of the parallel scan chains is bypassed as shown in Figure 102 (page 212). It is possible to concatenate the parallel scan chains from one core with the parallel scan chains of another in an SOC to reduce the size of the TAM as shown in Figure 31. Core1 has four parallel scan chains and Core2 has three parallel scan chains. If Core1 is under test, the parallel scan chains in Core2 can be put into bypass mode. In this way, the TAM can be kept smaller, while Core1 and Core2 can be tested separately, providing both cores have a WP_BYPASS instruction.

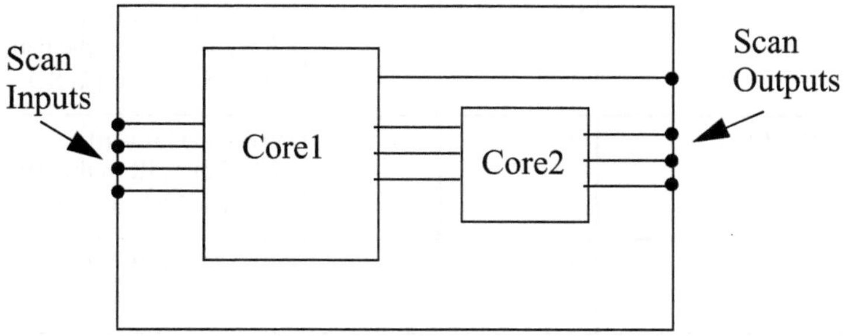

Figure 31 Utilizing WP_BYPASS

Table 7 EX Core instruction definitions

Instruction	*Mandatory*	*Standard*	*Description*
WS_BYPASS	Y	Y	•WBR in transparent mode •WBY in the path between WSI and WSO
WS_EXTEST	Y	Y	•WBR in Outward Facing mode •WBR in the path between WSI and WSO
WP_INTEST	Y	Y	•WBR in Inward Facing mode •WBY or WIR in the path between WSI and WSO
WS_INTEST	N	Y	•WBR in Inward Facing mode •WBR in the path between WSI and WSO
WP_INTEST_MBIST	N	N	•WBY or WIR in the path between WSI and WSO
WP_EXTEST	N	Y	•WBR in Outward Facing mode •WBY or WIR in the path between WSI and WSO
WS_SAFE_SINGLE	N	N	•WBY in the path between WSI and WSO
WP_INTEST_SEQ	N	N	•WBR in Inward Facing mode •WBY or WIR in the path between WSI and WSO
WP_EXTEST_SEQ	N	N	•WBR in Outward Facing mode •WBY or WIR in the path between WSI and WSO
WP_BYPASS	N	N	•WBY or WIR in the path between WSI and WSO

For each instruction, the WBR and WBY must be configured through static test signals. In addition, any static signals that control the mode of the core must also be controlled to the proper state for that instruction through the WIR.

5.3 CTL Template for 1500 Instructions

In previous chapters the beginning of a CTL code containing a declaration statement for all wrapper level signals were defined. Also, a **CTLMode** block defining wrapper interface port terminals along with the timing parameters required to operate these terminals was defined. The **CTLMode** block discussed then was particular in that it was nameless so as to provide inheritance to other, named, **CTLMode** blocks. In the following paragraphs use of this inheritance feature will complete the definition of test modes in CTL as they relate to 1500 instructions.

The purpose of a 1500 instruction is to configure the 1500 wrapper so that a particular test can be applied to this wrapper. This definition is identical to that of a test mode as discussed in chapter 4. Therefore, 1500 instructions are not different from CTL test modes and as such will be described using the **CTLMode** keyword as represented below:

```
<5.1>   Environment (ENV_NAME) {
<5.2>           (CTLMode (CTLMODE_NAME) {
<5.3>           } )*
<5.4>   }
```

The following sections will present the various CTL components required to complete the definition of a CTL template for 1500 instructions.

5.3.1 Describing Patterns in CTL

As presented earlier, compliance with 1500 relies on usage of 1450.6 (CTL) syntax to express wrapper logic and pattern information. Understanding how to express patterns in CTL is therefore an essential condition to achieving 1500 compliance. This section will cover basic CTL syntax for description of test patterns.

Due to the syntax proximity between CTL and STIL, a CTL pattern syntax is based on STIL pattern constructs but it should be noted that CTL does have specificities that would make some STIL pattern syntax invalid in CTL. The following paragraphs will cover CTL syntax that allows definition of pattern data and pattern protocol.

5.3.1.1 Describing Pattern Data in CTL

1450.6 defines the following construct for the specification of pattern data:

```
<5.5>   Pattern PATTERN_NAME {
<5.6>        ( Setup { (sigref_expr = value_variable_expr;)* })
<5.7>        <((LABEL:) P;)* |
<5.8>        ((LABEL:) P { (sigref_expr = value_variable_expr;)*})*>
<5.9>        // Pattern statements that are allowed from the statements defined in
<5.10>       // IEEE Std. 1450-1999 and 1450.1
<5.11>       ((LABEL:) Loop LOOPCNT { (P-statements)* })*
<5.12>       ((LABEL:) Goto LABELNAME;)*
<5.13>       ((LABEL:) BreakPoint;)*
<5.14>       ((LABEL:) X TAG;)*
<5.15>  }
```

The above **Pattern** construct allows for description of a variety of pattern data types. However, for our discussion we will limit the pattern block syntax to the following subset:

```
<5.16>  Pattern PATTERN_NAME {
<5.17>       ( Setup { (sigref_expr = value_variable_expr;)* })
<5.18>       P { (sigref_expr = value_variable_expr;)*}
<5.19>  }
```

where:

- **Setup** represents a call to a setup or initialization macro meant to be executed prior to the test data represented with P{ } statements.
- **P** represents a call to the protocol procedure or protocol macro associated with PATTERN_NAME. The syntax to be used to describe protocols is discussed in Section *5.3.1.1.1* (see page 95).
- *sigref_expr* represents the name of the terminal where pattern data is being applied and *value_variable_expr* represents a stream of values that constitutes the pattern data.

The definition of **Pattern** block keywords that are not discussed here can be found in the 1450.6 document.

The **Pattern** block is defined outside the **Environment** block with the following CTL code structure:

```
Signals { }
SignalGroups { }

Environment {
    CTLMode { }
}
(Pattern PATTERN_NAME {
    P { (sigref_expr = value_variable_expr;)*}
})*
```

Patterns defined in the **Pattern** block will be referred to in a **CTLMode** block which will specifically define the purpose of the patterns in the context of the test mode associated with this **CTLMode** block. This will be discussed in the following paragraphs.

5.3.1.1.1 Describing Pattern Protocol

In CTL, protocols are defined in **MacroDefs** blocks using macro and procedure constructs. These macros and procedures resemble programming language functions which contain a pattern protocol, and will get called with arguments that constitute pattern data. Therefore, call and execution of macros or procedures causes pattern data to be applied according to the protocol defined inside the macros or procedures. **MacroDefs** blocks are defined outside the **Environment** block with the following CTL code structure:

```
Signals { }
SignalGroups { }

MacroDefs (MACRO_DOMAIN_NAME) {
    ( MACRO_NAME {
        (PATTERN_STATEMENT)*
    })*
}
Environment {
    CTLMode { }
}
Pattern PATTERN_NAME {
    P { (sigref_expr = value_variable_expr;)*}
}
```

Where:

- MACRO_DOMAIN_NAME represents the name given to the group of macros defined in the **MacroDefs** block. This name is optional but it should be noted that nameless macro domains get inherited in a way similar to the inheritance of the **CTLMode** block discussed in Section *4.2* (see page 38).

- MACRO_NAME represents the name of a macro
- PATTERN_STATEMENT describes the behavior of the macro

Macros defined in the **MacroDefs** block will be referred to in a **CTLMode** block which will specifically combine the corresponding protocol with the appropriate pattern data. A typical example of protocol and data combination in CTL is that of a shift operation into a shift register. Let's consider that a sequence of "001" needs to be shifted into a 3-bit register driven by the following signals:

- "Data" as data pin
- "clk" as clock pin

This would be achieved by the following macro:

```
<5.44>  MacroDefs {
<5.45>      ShiftIntoRegister {
<5.46>          W shift_waveform;
<5.47>          Shift {
<5.48>          V { "Data"=#; "clk"=P;}
<5.49>          }
<5.50>      }
<5.51>  }
```

The above is a nameless **MacroDefs** block where:

- ShiftIntoRegister is the name of a macro
- shift_waveform represents the timing waveform to be used to apply data during execution of the macro
- V { } is a vector statement with represents one timing period worth of activity on design terminals. Duration of this timing period is specified in a waveform definition that exists outside the **MacroDefs** block.
- "#" is a place-holder single-bit variable that will be substituted by the values (pattern data) provided in the argument of the macro call.

With the above macro, the following macro call statement will complete the shift example:

```
<5.52>  Macro ShiftIntoRegister {"Data"=001;}
```

The above line calls the macro by the name of ShiftIntoRegister with "001" as argument to be applied to the "Data" terminal. As a result, during application of the ShiftIntoRegister macro, each bit of the argument data (001) will substitute the "#" sign that is defined in the macro. The result is the following sequence which allows "001" to be shifted into the register example:

```
V { "Data"=0; "clk"=P;}
V { "Data"=0; "clk"=P;}
V { "Data"=1; "clk"=P;}
```

The concept of macro calls will be used throughout the following sections and is essential to understanding CTL code in general and therefore is essential to the goal of describing compliance to the 1500 standard.

The structural description of the CTL code written thus far is shown below. Detailed information was removed from the below CTL code so as to allow a concise review of the code.

```
STIL 1.0 {
Design 2005;
CTL 2005;
}
Signals {
}
SignalGroups {
}

MacroDefs (MACRO_DOMAIN_NAME) {
    ( MACRO_NAME {
        (PATTERN_STATEMENT)*
    })*
} // MacroDefs block
Environment EX_wrapper {
CTLMode {
        Internal {
            wsp_wrck { } // wsp_wrck block
            wpp_wrck { } // wpp_wrck block
            WRSTN { } // WRSTN block
            SelectWIR { } // SelectWIR block
            ShiftWR { } // ShiftWR block
            CaptureWR { } // CaptureWR block
            UpdateWR { } // UpdateWR block
            CLK { } // CLK block
            WSI { } // WSI block
            WSO { } // WSO block
            WPSI [0..3] { } // WPSI [0..3] block
```

```
<5.84>                WPSO[0..3] { } // WPSO[0..3] block
<5.85>                MBISTFAIL { } // MBISTFAIL block
<5.86>                MBISTDONE { } // MBISTDONE block
<5.87>                MBISTLOGOUT { } // MBISTLOGOUT block
<5.88>                WPSE { } // WPSE block
<5.89>                MBISTRUN { } // MBISTRUN block
<5.90>                MBISTLOG { } // MBISTLOG block
<5.91>            } // Internal block
<5.92>        } // CTLMode block
<5.93>    } // Environment block
<5.94>    Pattern PATTERN_NAME {
<5.95>        P { (sigref_expr = value_variable_expr;)*}
<5.96>    }
```

Because the above **CTLMode** block is nameless, the WSP and WPP information that it contains, as defined in Chapter 4, can be inherited as we define new test modes corresponding to the various 1500 instructions. Therefore, we will not repeat the **Internal** block description of 1500 terminals inside each test mode as we proceed.

The following paragraphs will describe CTL constructs that reside inside the **CTLMode** block and allow reference to pattern data. These keywords are:

- **DomainReferences**
- **TestMode**
- **PatternInformation**
- **TestModeForWrapper**

The corresponding CTL code structure follows:

```
<5.97>    Environment (ENV_NAME) {
<5.98>        (CTLMode (CTLMODE_NAME) {
<5.99>            ( TestMode (test_mode_enum)+ ; )
<5.100>           ( DomainReferences ( CORE_INSTANCE_NAME)* { })*
<5.101>           (Internal { } )
<5.102>           (PatternInformation { } )
<5.103>       (TestModeForWrapper WRAPPER_TEST_MODE
              TEST_MODE_CODE; )
<5.104>       } )* CTLMode block
<5.105>   } // Environment block
```

5.3.2 Accessing Information Outside the CTLMode Block Using the DomainReferences Keyword

Not all information described outside the **Environment** block is visible to the **Environment** block. In general, only nameless STIL blocks are inherited by the **Environment** block. As a result, CTL provides a construct for making visible information (such as pattern data and pattern protocol) that is defined outside the **Environment** block but is required by the current **CTLMode** block. This construct is the **DomainReferences** construct defined in the **CTLMode** block. The **DomainReferences** keyword has the following syntax:

```
<5.106>    (DomainReferences ( CORE_INSTANCE_NAME)* {
<5.107>        (Category (CATEGORY_NAME)+; )
<5.108>        (DCLevels (DC_LEVELS_NAME)+;)
<5.109>        (DCSets (DC_SETS_NAMES)+;)
<5.110>        (MacroDefs ( MACRO_DEF_NAME )+ ; )
<5.111>        (Procedures ( PROCEDURE_NAME )+ ; )
<5.112>        (Selector ( SELECTOR_NAME)+;)
<5.113>        (SignalGroups ( SIGNALGROUPS_NAME )+ ; )
<5.114>        (ScanStructures ( SCAN_STRUCT_NAME )+ ; )
<5.115>        (Timing ( TIMING_NAME )+ ; )
<5.116>        (Variables ( VARIABLES_NAME)+; )
<5.117>    } )* // DomainReferences block
```

For this discussion, the following **DomainReferences** keywords will be used:

- **MacroDefs** to refer to macros (defined outside the Environment block) used in patterns that accompany wrapped or unwrapped cores
- **ScanStructures** to refer to scan chains (defined outside the Environment block) and operated by the patterns provided for the unwrapped or wrapped core.

These keywords are STIL constructs that exist outside the **Environment** block and arguments corresponding to these keywords refer to the name given to these STIL constructs when they were defined.

The remaining **DomainReferences** keywords are beyond the scope of this discussion and will not be used here but can be found in the 1450.6 document.

5.3.3 Describing Test Modes Using the TestMode Keyword

1450.6 defines that a given **CTLMode** block must contain a **TestMode** statement that identifies the type of test mode being described, unless this **CTLMode** block is nameless. The CTL syntax to be used for this identification is the following:

TestMode < **Bypass** | **Controller** | **Debug** | **ExternalTest** | **ForInheritOnly** | **InternalTest** | **Isolate** | **Normal** | **PowerOff** | **PowerOn** | **PreLoad** | **Quiet** | **Repair** | **Sample** | **Transparent** | **User** USER_DEFINED >

For this discussion, the above syntax will be restricted to a subset of arguments that meet 1500 needs. The resulting syntax will be:

TestMode < **Bypass** | **ExternalTest** | **InternalTest** >

where:

- **Bypass** represents a test mode that uses the wrapper bypass register
- **ExternalTest** represents a test mode meant to test logic external to the core
- **InternalTest** identifies a test mode intended for logic inside the core being wrapped.

5.3.4 Specifying Pattern Information for Test Modes Using the PatternInformation Keyword

1500 does not codify usage of the **PatternInformation** CTL keyword. Information related to the syntax and usage of this keyword must therefore come from 1450.6. An overview of the 1450.6 **PatternInformation** syntax follows:

<5.118> **PatternInformation** {
<5.119> < **Pattern** | **PatternBurst** > *(pat_or_burst_name)* { }
<5.120> **PatternExec** (EXEC_NAME) { }
<5.121> <**Procedure** | **Macro** > PROCEDURE_OR_MACRO_NAME { }
<5.122> **WaveformTable** (WFT)* { }
<5.123> }

The **WaveformTable** keyword inside the **PatternInformation** block will not be used in this discussion. Instead, a **WaveformTable** block will be defined in the nameless CTL block so that this information can be inherited by all (named) 1500 instruction **CTLMode** blocks. Other keywords of the **PatternInformation** block will be discussed individually in the following paragraphs.

5.3.4.1 The Pattern and PatternBurst Constructs

The difference between **Pattern** and **PatternBurst** is that **Pattern** refers to a sequence of test data while **PatternBurst** defines an ordered sequence of patterns. 1450.6 defines the **Pattern** and **PatternBurst** syntax as follows:

```
<5.124>   < Pattern | PatternBurst > (pat_or_burst_name) {
<5.125>      (Purpose (pattern_or_burst_enum)+ ;)
<5.126>      ( CoreInstance (CORE_INSTANCE_NAME)+; )
<5.127>      ( CycleCount integer;)
<5.128>      ( Power power_expr < Average | Maximum > ; )*
<5.129>      ( Fault { } )
<5.130>      ( FileName FILE_NAME ;)
<5.131>      ( ForeignPatterns {} )
<5.132>      ( Identifiers { } )
<5.133>      ( Protocol <Macro | Procedure> NAME (SETUP_NAME); )
<5.134>   } )* // Pattern | PatternBurst block
```

Let's restrict the above syntax to keywords required for our discussion. Information about removed keywords can be obtained from 1450.6 document. However, before doing this it should be noted that while all of these keywords are valid CTL keywords, 1500 does not allow use of the **ForeignPatterns** block to represent patterns. The purpose of the **ForeignPatterns** construct is to allow user-selected pattern formats to be linked to the CTL code. The reason behind the 1500 restriction is that **ForeignPatterns** (or non-CTL-native) pattern formats do not necessarily satisfy the requirement for data and protocol separation. As established in Section *1.3.1* (see page 9), this separation is a condition to efficient core-level to SOC-level pattern porting. This 1500 restriction on use of the **ForeignPatterns** construct establishes the notion of a 1500 compliant CTL, in that a 1450.6 compliant CTL code is not necessarily 1500 compliant.

For this discussion the **Pattern** and **PatternBurst** keywords will be limited to the following syntax:

```
<5.135>   Pattern | PatternBurst (pat_or_burst_name) {
<5.136>   Purpose (pattern_or_burst_enum)+
<5.137>   Protocol <Macro | Procedure> NAME (SETUP_NAME); )
<5.138>   }
```

Where:

- **Pattern** or **PatternBurst** identifies pattern data *(pat_or_burst_name)* associated with the test mode being defined.

- **Purpose** defines the role played by the pattern specified in the test mode being defined. 1450.6 defines possible **Purpose** values as follows:
  <AtSpeed | ChainContinuity | CompatibilityInformation | Endurance | EstablishMode | IDDQ | LogicBIST | MemoryBIST | Padding | Parametric | Retention | Scan | TerminateMode | User USER_DEFINED>. 1500 mandates separation of patterns that are meant to establish a test mode from patterns that are meant for execution after the appropriate mode has been established. Use of the **Purpose** construct allows distinction between these two types of patterns. Further description of these **Purpose** values is available in the 1450.6 document.
- **Protocol** defines the protocol corresponding to the pattern data, and SETUP_NAME represents the name of a setup macro or procedure that is to be invoked at the beginning of the application of the pattern data.

5.3.4.2 The PatternExec Construct

PatternExec identifies a standalone pattern structure where all information required to run a complete test is specified by a call to all **Pattern** and **PatternBurst** constructs that make up the standalone pattern set. 1450.6 defines the **PatternExec** syntax as follows:

```
<5.139>   ( PatternExec (EXEC_NAME) {
<5.140>      (Purpose (EXEC_ENUM)+ ;)
<5.141>      ( PatternBurst (BURST_NAME)+ ; )
<5.142>      ( CycleCount integer;)
<5.143>      ( Power power_expr < Average | Maximum > ; )*
<5.144>      ( Fault { } )*
<5.145>   } )* // PatternExec block
<5.146>
```

Out of the above syntax let's eliminate the optional **Fault** construct (used for fault accounting) the optional **Power** construct (used to quantify power dissipation) and the **CycleCount** construct used to indicate the length of the patterns in cycle count. The following syntax remains:

```
<5.147>   PatternExec (EXEC_NAME) {
<5.148>      (Purpose (EXEC_ENUM)+ ;)
<5.149>      ( PatternBurst (BURST_NAME)+ ; )
<5.150>   }
```

Where

- EXEC_NAME refers to a **PatternExec** block name defined outside the **Environment** block.
- The **Purpose** keyword specifies the purpose of the **PatternExec** sequence. Valid arguments for the **Purpose** keyword are **Diagnostics**, **Production**, **Characterization** and **Verification**. These are "exec_enum" values defined by 1450.6.
- The **PatternBurst** keyword refers to the top-most burst or sequence of patterns that is defined for EXEC_NAME

5.3.4.3 The Procedure and Macro Constructs

The **Procedure** and **Macro** constructs declare protocols that are used by the patterns indicated in the **Pattern** construct discussed in Section *5.3.4.1* (see page 100). The **Procedure** or **Macro** keyword syntax follows:

```
<5.151>  <Procedure | Macro > PROCEDURE_OR_MACRO_NAME {
<5.152>     ( Purpose ( procedure_or_macro_enum )+ ;)
<5.153>     ( ScanChain (CHAINNAME)+;)*
<5.154>     ( UseByPattern (pattern_or_burst_enum)+; )
<5.155>     ( Identifiers { } )*
<5.156>  } // Procedure | Macro block
```

Where:

- **Procedure** or **Macro** identifies the name of the procedure or macro that defines the protocol to be used to run the patterns specified in Section *5.3.4.1* (see page 100).
- **Purpose** defines the objective of the procedure or macro that is being defined. Possible **Purpose** values are defined by 1450.6 as: < **Capture** | **Control** | **DoTest** | **DoTestOverlap** | **Hold** | **Instruction** | **MemoryPrecharge** | **MemoryRead** | **MemoryReadModifyWrite** | **MemoryRefresh** | **MemoryWrite** | **ModeControl** | **Observe** | **Operate** | **ShiftIn** | **ShiftOut** | **Transfer** | **Update** | **User** USER_DEFINED >.
- **UseByPattern** specifies the use of the macro by the calling pattern. Possible uses are predefined by 1450.6 in the *pattern_or_burst_enum* list which comprises: < **AtSpeed** | **ChainContinuity** | **CompatibilityInformation** | **Endurance** | **EstablishMode** | **IDDQ** | **LogicBIST** | **MemoryBIST** | **Padding** | **Parametric** | **Retention** | **Scan** | **TerminateMode** | **User** USER_DEFINED >.

The **Identifiers** keyword will not be needed in this discussion. Explanation about this keyword can be obtained in the 1450.6 document.

5.3.5 Identifying IEEE 1500 Instructions Using the TestModeForWrapper Keyword

The **TestModeForWrapper** keyword serves to identify 1500 instructions or test modes by name. The **TestModeForWrapper** syntax follows:

(**TestModeForWrapper** WRAPPER_TEST_MODE TEST_MODE_CODE*;* *)*

Where:

- WRAPPER_TEST_MODE corresponds to the reserved 1500 instruction name that matches the test mode being defined in the current Test Mode block. Example names would be WS_EXTEST, WP_INTEST and WS_BYPASS
- TEST_MODE_CODE represents the unique opcode that is defined to identify the instruction being described as required by 1500

5.4 Combined 1500 Instructions CTL Code Template

The following CTL code template results from the above paragraphs and will serve as description vehicle for 1500 instructions defined for the EX wrapper in Section *5.2* (see page 89).

```
STIL 1.0 {
Design 2005;
CTL 2005;
}

Signals {
// Signal list
}
SignalGroups {
// Signal group list
}

Environment EX_wrapper {
    CTLMode {
```

```
<5.171>         Internal {
<5.172>             wsp_wrck { } // wsp_wrck block
<5.173>             wpp_wrck { } // wpp_wrck block
<5.174>             WRSTN { } // WRSTN block
<5.175>             SelectWIR { } // SelectWIR block
<5.176>             ShiftWR { } // ShiftWR block
<5.177>             CaptureWR { } // CaptureWR block
<5.178>             UpdateWR { } // UpdateWR block
<5.179>             CLK { } // CLK block
<5.180>             WSI { } // WSI block
<5.181>             WSO { } // WSO block
<5.182>             WPSI [0..3] { } // WPSI [0..3] block
<5.183>             WPSO[0..3] { } // WPSO[0..3] block
<5.184>             MBISTFAIL { } // MBISTFAIL block
<5.185>             MBISTDONE { } // MBISTDONE block
<5.186>             MBISTLOGOUT { } // MBISTLOGOUT block
<5.187>             WPSE { } // WPSE block
<5.188>             MBISTRUN { } // MBISTRUN block
<5.189>             MBISTLOG { } // MBISTLOG block
<5.190>         } // Internal block
<5.191>     } // CTLMode block
<5.192>     (CTLMode (CTLMODE_NAME) {
<5.193>         (TestMode (test_mode_enum)+ ; )
<5.194>         (Internal { } )
<5.195>         (PatternInformation { } )
<5.196>         TestModeForWrapper TEST_MODE TEST_MODE_CODE;
<5.197>     } )* // CTLMode (CTLMODE_NAME) block
<5.198> } // Environment block
```

Lines 5.162 through 5.191 are fully defined in Section *4.4* (see page 68).

Note that, so far, only 1500 terminals have been specified inside the Internal block of the nameless CTLMode block. Functional EX wrapper terminals will be specified locally to the CTLMode block corresponding to the EX wrapper instructions.

5.5 Describing CTL Code for EX Wrapper Instructions

In the following sections the 1500 instruction CTL code syntax discussed in Section *5.4* (see page 104) will be utilized to describe CTL code corresponding to each EX wrapper instructions. The EX wrapper instruction opcodes defined in Section *8.3.2* (see page 198) are repeated in Table 8 for convenience.

Table 8 EX wrapper instruction opcodes

instruction	*opcode*
WS_BYPASS	0000
WS_EXTEST	0001
WS_INTEST	0010
WS_SAFE_SINGLE	0011
WP_INTEST	0100
WP_BYPASS	0101
WP_EXTEST	0110
WP_INTEST_MBIST	0111
WP_EXTEST_SEQ	1001
WP_INTEST_SEQ	1010

5.5.1 Describing WS_BYPASS in CTL for the EX Wrapper

Understanding the sequence of events involved in the loading of the WS_BYPASS instruction is a pre-requisite to describing WS_BYPASS in CTL. We will therefore start by describing the WS_BYPASS operation.

5.5.1.1 Loading the WS_BYPASS Instruction

A specific sequence of operations must occur in order to load the WS_BYPASS instruction into the EX wrapper. This sequence of operations comprises the following steps which assume that the wrapper has been placed into "wrapper enabled" state by a prior operation:

1. Select the WIR by applying "1" on the SelectWIR terminal
2. Enable shift of data into the wrapper by applying "1" on the ShiftWR terminal.
3. Apply the WS_BYPASS opcode to WSI and use WRCK to shift the opcode into the WIR
4. De-activate ShiftWR when shift of WS_BYPASS opcode is complete. This will prevent subsequent rising edges of WRCK from corrupting data existing in the shift stage of the WIR.
5. Activate the WIR operation by applying "1" on the UpdateWR terminal. This will make the WS_BYPASS instruction available for decod-

ing and will generate the appropriate control signals. These control signals will put the various wrapper registers and the core into the appropriate configuration for the bypass mode.

6. Deselect the WIR by applying "0" on the SelectWIR terminal. This will keep the loaded instruction active.

The above sequence of operations is described in Figure 32 The wrapper reset operation was excluded from the above sequence so as to make this sequence generic enough to accommodate cases where consecutive load of different instructions is required. In such cases a wrapper reset between instructions would not be desirable.

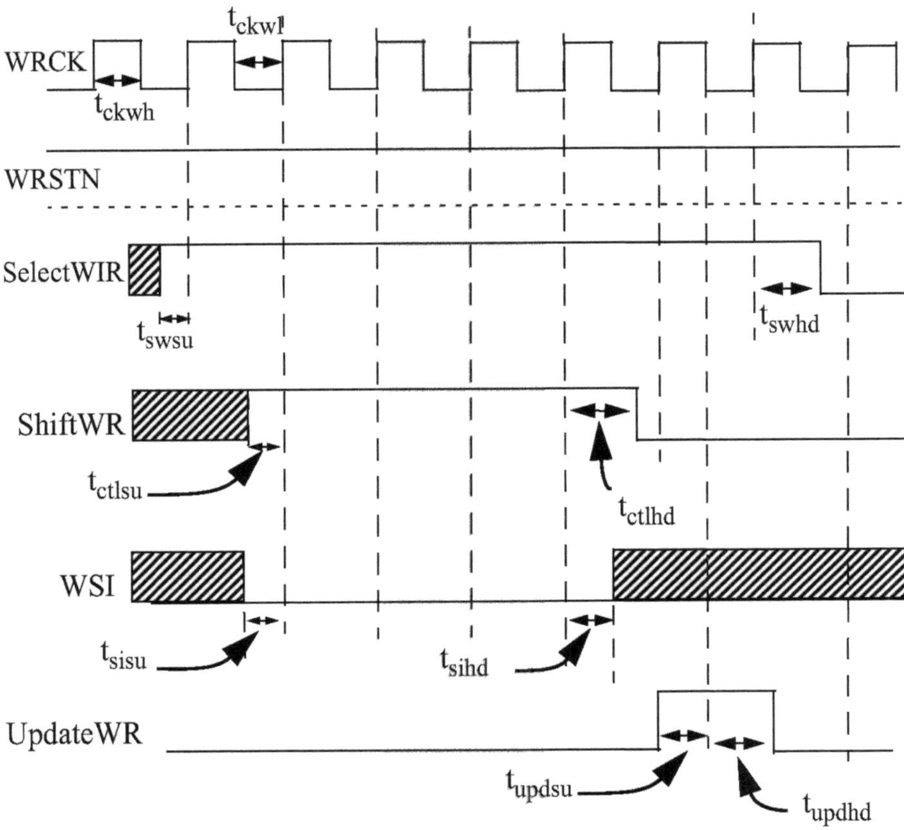

Figure 32 EX wrapper WS_BYPASS instruction load timing

Note: WS_BYPASS is required to be active in the WIR as a result of a wrapper reset operation. This is shown in Figure 85 (page 189). The alternate method shown here allows WS_BYPASS to be loaded without going through a wrapper reset operation.

5.5.1.2 Understanding WS_BYPASS Setup Protocol and Data

Describing the WS_BYPASS instruction in CTL is equivalent to describing Figure 32 in CTL. First let's separate protocol and data from this figure. While activity on all terminals described in Figure 32 is important, values placed on WSI are critical as these values contain the opcode of the WS_BYPASS instruction and the sole purpose of the sequence of operations visible in this figure is to allow movement of these values from WSI into the WIR. In Section *1.3.1* (see page 9), a protocol was defined as a sequence of events that enable application of test data. This definition allows us to recognize values on the WSI terminal as data and values on WRCK, WRSTN,

SelectWIR, ShiftWR, and UpdateWR as protocol. In other words, we have just described the two components (protocol and data) of a pattern where the WS_BYPASS opcode (0000) constitutes test data and value on the WSP terminals listed above constitute test protocol. This separation between data and protocol is what will ultimately allow the protocol part of the pattern to be adapted to the SOC environment so that the WS_BYPASS opcode can be loaded into the wrapped core from the SOC level.

5.5.1.3 Describing the EX Wrapper WS_BYPASS Setup Pattern in CTL

In the previous sections a pattern was identified that establishes the WS_BYPASS mode of the EX wrapper. This pattern will be described in terms of macro and pattern data with the syntax established in Section *5.3.1* (see page 93). We will name the corresponding macro "instruction_mode_setup". The resulting CTL code follows:

```
<5.199>  MacroDefs ex_wrapper_macros {
<5.200>     instruction_mode_setup {
<5.201>     W WSP_WPP_waveform;
<5.202>     V {"WRCK"=P; "WRSTN"=1; "UpdateWR"=0;}
<5.203>     V {"SelectWIR"=1;}
<5.204>     C {"ShiftWR"=1; }
<5.205>     Shift {
<5.206>        V { "WSI"=#; "WRCK"=P;}
<5.207>     }
<5.208>     V { "WRCK"=P; "ShiftWR"=0;}
<5.209>     V {"UpdateWR"=1;}
<5.210>     V {"UpdateWR"=0;{
<5.211>     V {"SelectWIR"=0;}
<5.212>     } // instruction_mode_setup block
<5.213>  } // MacroDefs block
```

The C { } statement is a STIL construct that defines constraints that are to be applied in the following cycle (or V{ } statement) i.e. the first shift cycle in this case. This was added to match the waveform shown in Figure 32 (page 108) where ShiftWR is activated concurrently with the application of data on the WSI terminal.

"WSP_WPP_waveform" is the timing waveform for application of WSP and WPP terminals. As an example timing, this waveform will define that data is applied on all WSP and WPP terminals, with the exception of clock terminals, at time "0" of the test cycle. The WRCK clock edges will be defined at 45ns and 55ns respectively in a 100ns test cycle. The resulting waveform timing follows.

```
Timing {
    WaveformTable WSP_WPP_waveform {
        Period '100ns';
        Waveforms {
            "WRSTN" { 01X { '0ns' D/U/N; }}
            "SelectWIR" { 01X { '0ns' D/U/N; }}
            "ShiftWR" { 01X { '0ns' D/U/N; }}
            "CaptureWR" { 01X { '0ns' D/U/N; }}
            "UpdateWR" { 01X { '0ns' D/U/N; }}
            'WSI+WPSI[0..3]' { 01X { '0ns' D/U/N; }}
            "WPSE" { 01X { '0ns' D/U/N; }}
            'MBISTRUN+MBISTLOG' { 01X { '0ns' D/U/N; }}
            'CLK+WRCK' { P { '0ns' D; '45ns' U; '55ns' D; }}
            'WSO+WPSO[0..3]' { { '10ns' lhXt; }}
            'MBISTFAIL+MBISTDONE+MBISTLOGOUT' { { '50ns' lhXt; }}
        } // Waveforms block
    } // WaveformTable block
}// Timing block
```

The pattern data using the above bypass macro would be:

```
Pattern WsBypassSetupPattern {
    P {"WSI"=0000;}
}
```

5.5.1.4 Defining the CTLMode Block for the WS_BYPASS Mode

The below CTL code is assumed preceded by the signal group names and the nameless **CTLMode** block defined in Section *4.4* (see page 68). This nameless **CTLMode** block defines information that is to be inherited by the below (named) **CTLMode** block. The "EX_wrapper_WSP_chains" **ScanStructures** block used by the below CTL code is defined in Section *8.4* (see page 205). For this discussion only the WIR chain will be needed to load the instruction opcode.

```
CTLMode WS_BYPASS_MODE {
    TestMode Bypass;
    DomainReferences {
        MacroDefs ex_wrapper_macros;
        ScanStructures EX_wrapper_WSP_chains;
    } // DomainReferences block
    Internal {
        all_functional {DataType Functional;}
        wpp_mbist {DataType Unused;}
        wpp_scan {DataType Unused;}
    } // Internal block
```

```
    PatternInformation {
        Pattern WsBypassSetupPattern {
            Purpose EstablishMode;
            Protocol Macro instruction_mode_setup;
        } // Pattern block
        Macro instruction_mode_setup {
            Purpose ModeControl;
            UseByPattern EstablishMode;
        } // Macro block
    } // PatternInformation block
    TestModeForWrapper WS_BYPASS 0000;
} // CTLMode WS_BYPASS_MODE block
```

The above CTL code defines enough information to establish the WS_BYPASS test mode but because of the type of configuration being described, this CTL code does not contain information about patterns to be executed once the mode has been established.

In the following sections, knowledge acquired during creation of the CTL code for the bypass instruction will be used to define CTL code for the remaining EX wrapper instructions.

5.5.2 Describing WP_BYPASS in CTL for the EX Core

Load of the WP_BYPASS instruction is similar to the load of the WS_BYPASS instruction discussed in Section *5.5.1.1* (see page 106) except that data on the WSI terminal will be specific to the WP_BYPASS instruction. The corresponding timing diagram follows in Figure 33.

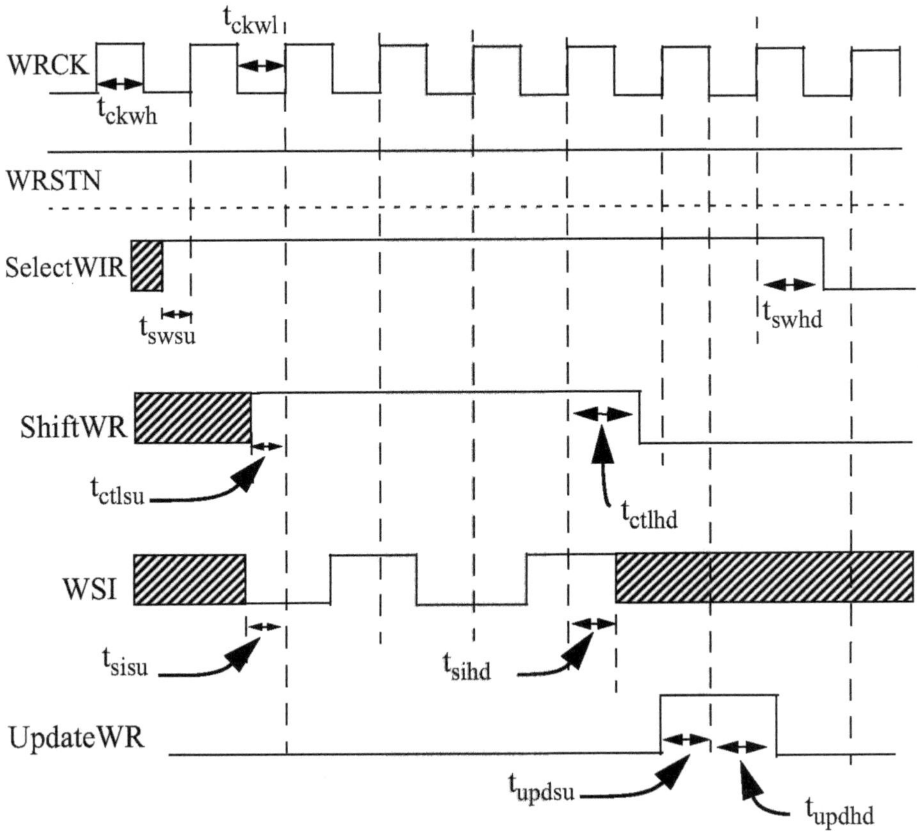

Figure 33 EX wrapper WP_BYPASS instruction load timing

5.5.2.1 Defining the CTLMode block for the WP_BYPASS Mode

In this section the **MacroDefs** block defined in Section *5.5.1.3* (see page 109) will be reused. The pattern utilized to set up the WP_BYPASS instruction follows:

```
<5.258>   Pattern WpBypassSetupPattern {
<5.259>       P {"WSI"=0101;}
<5.260>   }
```

The below CTL code is assumed preceded by the signal group names and nameless **CTLMode** block defined in Section *4.4* (see page 68). This nameless **CTLMode** block defines information that is to be inherited by the below (named) **CTLMode** block. The "EX_wrapper_WSP_chains" **ScanStruc-**

tures block used by the below CTL code is defined in Section *8.4* (see page 205).

```
<5.261>  CTLMode WP_BYPASS_MODE {
<5.262>     TestMode Bypass;
<5.263>     DomainReferences {
<5.264>        MacroDefs ex_wrapper_macros;
<5.265>        ScanStructures EX_wrapper_WSP_chains
                              EX_wrapper_WPP_chains;
<5.266>     } // DomainReferences block
<5.267>     Internal {
<5.268>        all_functional {DataType Functional;}
<5.269>        wpp_mbist {DataType Unused;}
<5.270>     } // Internal block
<5.271>     PatternInformation {
<5.272>        Pattern WpBypassSetupPattern {
<5.273>           Purpose EstablishMode;
<5.274>           Protocol Macro instruction_mode_setup;
<5.275>        } // Pattern block
<5.276>        Macro instruction_mode_setup {
<5.277>           Purpose ModeControl;
<5.278>           UseByPattern EstablishMode;
<5.279>        } // Macro block
<5.280>     } // PatternInformation block
<5.281>     TestModeForWrapper WP_BYPASS 0101;
<5.282>  } CTLMode WP_BYPASS_MODE block
```

The above CTL code defines enough information to establish the WP_BYPASS test mode but does not contain information about patterns to be executed once the mode has been established.

5.5.3 Describing WS_EXTEST in CTL for the EX Core

Load of the WS_EXTEST instruction is similar to the load of the WS_BYPASS instruction discussed in Section *5.5.1.1* (see page 106) except that data on the WSI terminal will be specific to the WS_EXTEST instruction. The corresponding timing diagram follows in Figure 34.

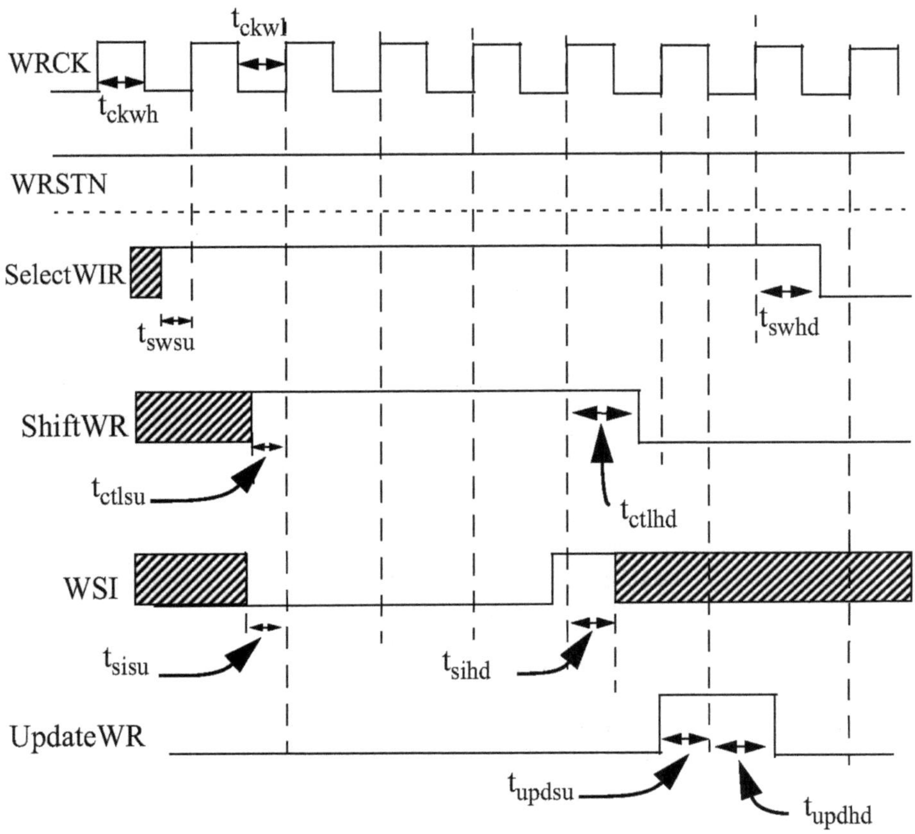

Figure 34 EX wrapper WS_EXTEST instruction load timing

5.5.3.1 Defining the CTLMode Block for the WS_EXTEST Mode

In this section the **MacroDefs** block defined in Section *5.5.1.3* (see page 109) will be reused. The pattern utilized to set up the WS_EXTEST instruction follows:

```
<5.283>   Pattern WsExtestSetupPattern {
<5.284>      P {"WSI"=0001;}
<5.285>   }
```

The following CTL code is assumed preceded by the signal group names and nameless **CTLMode** block defined in Section *4.4* (see page 68). This nameless **CTLMode** block defines information that is to be inherited by the below (named) **CTLMode** block. The "EX_wrapper_WSP_chains" **Scan-**

Structures block used by the below CTL code is defined in Section *8.4* (see page 205).

```
CTLMode WS_EXTEST_MODE {
    TestMode ExternalTest;
    DomainReferences {
        MacroDefs ex_wrapper_macros;
    ScanStructures EX_wrapper_WSP_chains;
    } // DomainReferences block
    Internal {
        all_functional {DataType Functional;}
        wpp_mbist {DataType Unused;}
        wpp_scan {DataType Unused;
    } // Internal block
    PatternInformation {
        Pattern WsExtestSetupPattern {
            Purpose EstablishMode;
            Protocol Macro instruction_mode_setup;
        } // Pattern block
        Macro instruction_mode_setup {
            Purpose ModeControl;
            UseByPattern EstablishMode;
        } // Macro block
    } // PatternInformation block
    TestModeForWrapper WS_EXTEST 0001;
} // CTLMode WS_EXTEST_MODE block
```

The above CTL code defines enough information to establish the WS_EXTEST test mode but does not contain information about patterns to be executed once the mode has been established.

5.5.4 Describing WP_INTEST in CTL for the EX core

Load of the WP_INTEST instruction is similar to the load of the WS_BYPASS instruction discussed in Section *5.5.1.1* (see page 106) except that data on the WSI terminal will be specific to the WP_INTEST instruction. The corresponding timing diagram follows in Figure 35.

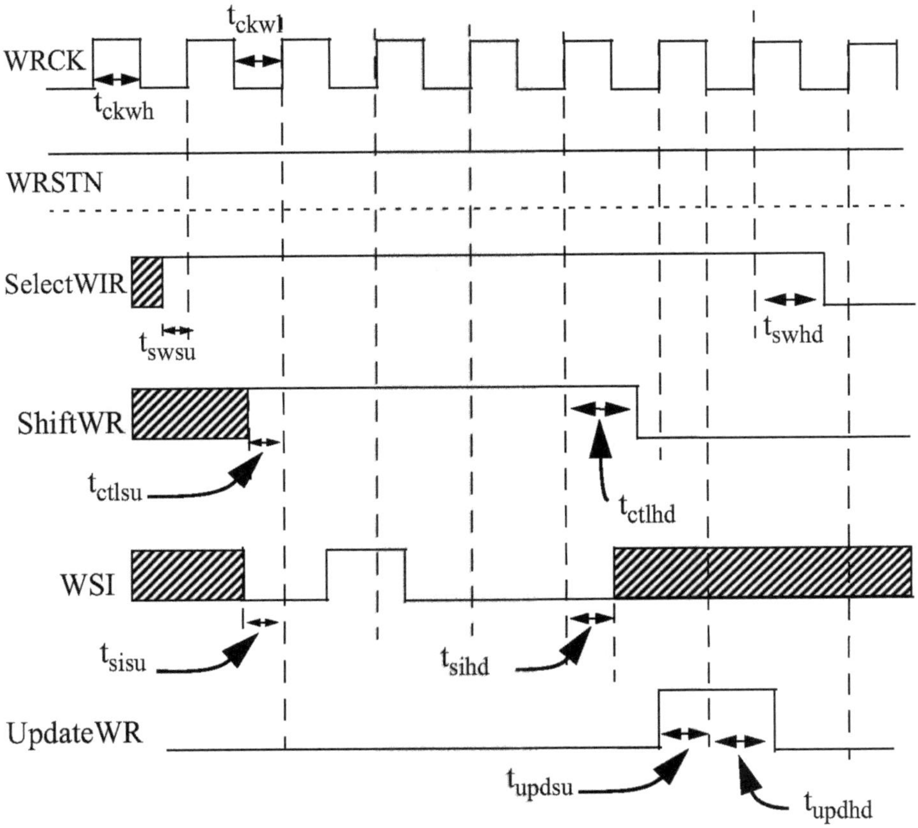

Figure 35 EX wrapper WP_INTEST instruction load timing

5.5.4.1 Defining the CTLMode Block for the WP_INTEST Mode

In this section the **MacroDefs** block defined in Section *5.5.1.3* (see page 109) will be reused. The pattern utilized to set up the WP_INTEST instruction follows:

```
<5.309>  Pattern WpIntestSetupPattern {
<5.310>      P {"WSI"=0100;}
<5.311>  }
```

The following CTL code is assumed preceded by the signal group names and nameless **CTLMode** block defined in Section *4.4* (see page 68). This nameless **CTLMode** block defines information that is to be inherited by the below (named) **CTLMode** block. The "EX_wrapper_WSP_chains" **Scan-**

Structures block used by the below CTL code is defined in Section *8.4* (see page 205).

```
<5.312>  CTLMode WP_INTEST_MODE {
<5.313>      TestMode InternalTest;
<5.314>          DomainReferences {
<5.315>              MacroDefs ex_wrapper_macros;
<5.316>              ScanStructures EX_wrapper_WSP_chains
                                    EX_wrapper_WPP_chains;
<5.317>          } // DomainReferences block
<5.318>      Internal {
<5.319>          all_functional {DataType Functional;}
<5.320>          wpp_mbist {DataType Unused;}
<5.321>      } // Internal block
<5.322>      PatternInformation {
<5.323>          Pattern WpIntestSetupPattern {
<5.324>          Purpose EstablishMode;
<5.325>          Protocol Macro instruction_mode_setup;
<5.326>          } // Pattern block
<5.327>          Macro instruction_mode_setup {
<5.328>              Purpose ModeControl;
<5.329>              UseByPattern EstablishMode;
<5.330>          } // Macro block
<5.331>      } // PatternInformation block
<5.332>      TestModeForWrapper WP_INTEST 0100;
<5.333>  } // CTLMode WP_INTEST_MODE block
```

The above CTL code defines enough information to establish the WS_BYPASS test mode but does not contain information about patterns to be executed once the mode has been established.

5.5.5 Describing WS_INTEST in CTL for the EX Core

Load of the WS_INTEST instruction is similar to the load of the WS_BYPASS instruction discussed in Section *5.5.1.1* (see page 106) except that data on the WSI terminal will be specific to the WS_INTEST instruction. The corresponding timing diagram follows in Figure 36.

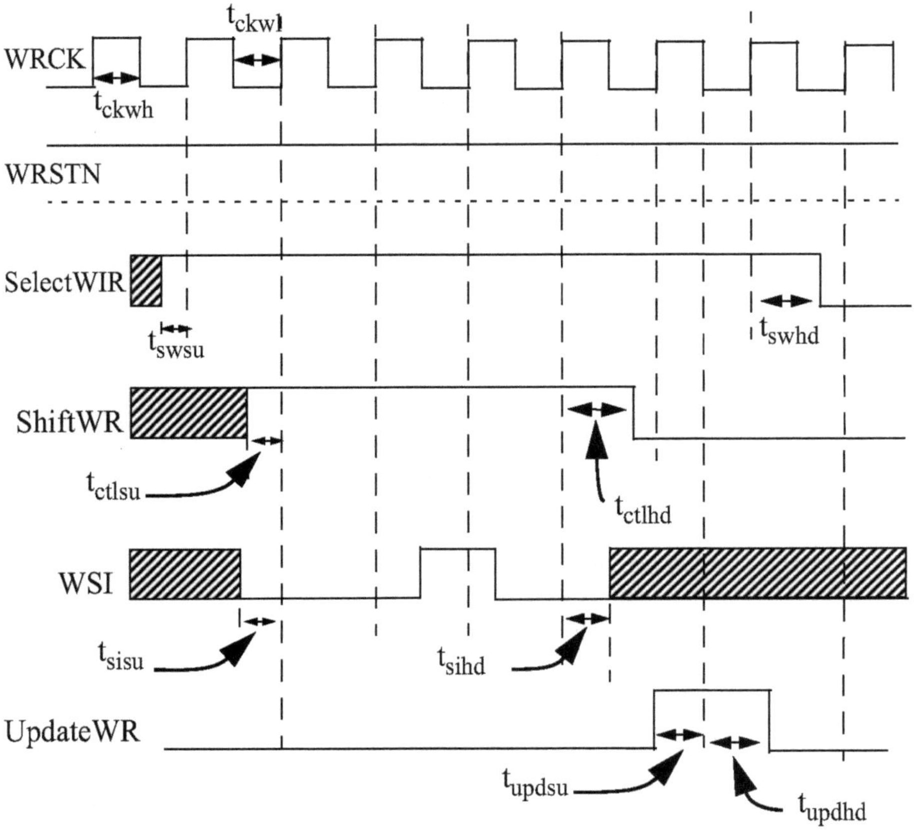

Figure 36 EX wrapper WS_INTEST instruction load timing

5.5.5.1 Defining the CTLMode Block for the WS_INTEST Mode

In this section the **MacroDefs** block defined in Section *5.5.1.3* (see page 109) will be reused. The pattern utilized to set up the WS_INTEST instruction follows:

```
<5.334>   Pattern WsIntestSetupPattern {
<5.335>      P {"WSI"=0010;}
<5.336>   }
```

The following CTL code is assumed preceded by the signal group names and nameless **CTLMode** block defined in Section *4.4* (see page 68). This nameless **CTLMode** block defines information that is to be inherited by the below (named) **CTLMode** block. The "EX_wrapper_WSP_chains" **Scan-**

Structures block used by the below CTL code is defined in Section *8.4* (see page 205).

```
CTLMode WS_INTEST_MODE {
    TestMode InternalTest;
    DomainReferences {
        MacroDefs ex_wrapper_macros;
        ScanStructures EX_wrapper_WSP_chains;
    } // DomainReferences block
    Internal {
        all_functional {DataType Functional;}
        wpp_mbist {DataType Unused;}
        wpp_scan {DataType Unused;
    } // Internal block
    PatternInformation {
        Pattern WsIntestSetupPattern {
            Purpose EstablishMode;
            Protocol Macro instruction_mode_setup;
        } // Pattern block
        Macro instruction_mode_setup {
            Purpose ModeControl;
            UseByPattern EstablishMode;
        } // Macro block
    } // PatternInformation block
    TestModeForWrapper WS_INTEST 0010;
} // CTLMode WS_INTEST_MODE block
```

The above CTL code defines enough information to establish the WS_INTEST test mode but does not contain information about patterns to be executed once the mode has been established.

5.5.6 Describing WP_INTEST_MBIST in CTL for the EX core

Load of the WP_INTEST_MBIST instruction is similar to the load of the WS_BYPASS instruction discussed in Section *5.5.1.1* (see page 106) except that data on the WSI terminal will be specific to the WP_INTEST_MBIST instruction. The corresponding timing diagram follows in Figure 37.

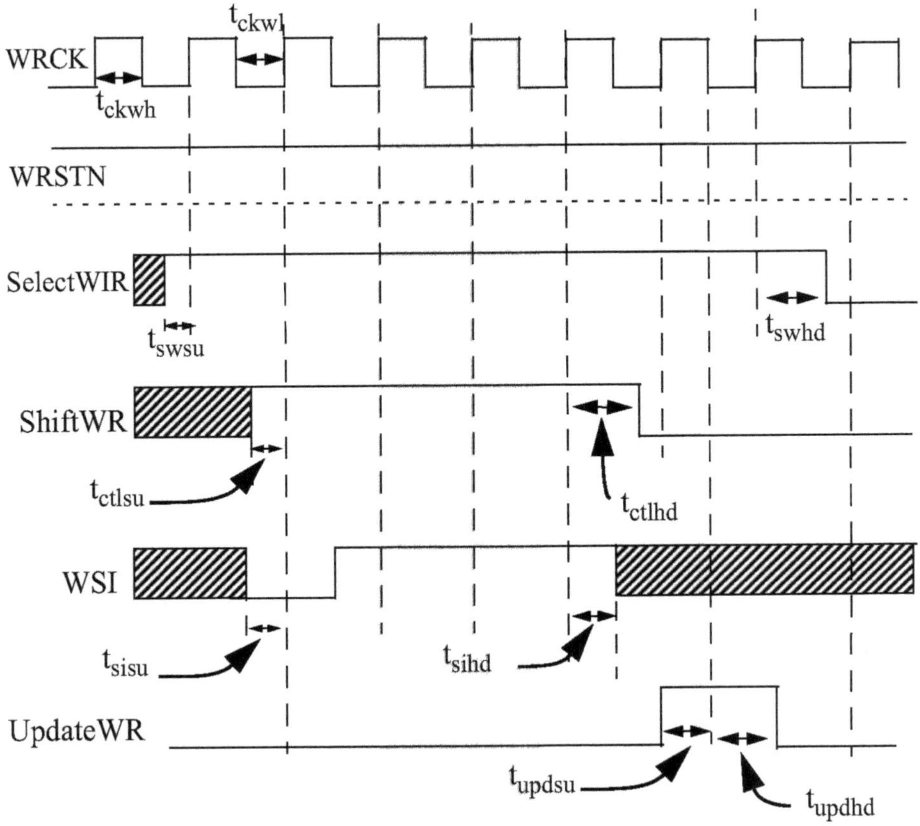

Figure 37 EX wrapper WP_INTEST_MBIST instruction load timing

5.5.6.1 Defining the CTLMode Block for the WP_INTEST_MBIST Mode

In this section the **MacroDefs** block defined in Section *5.5.1.3* (see page 109) will be reused. The pattern utilized to set up the WP_INTEST_MBIST instruction follows:

```
<5.360>   Pattern WpIntestMBistSetupPattern {
<5.361>      P {"WSI"=0111;}
<5.362>   }
```

The following CTL code is assumed preceded by the signal group names and nameless **CTLMode** block defined in Section *4.4* (see page 68). This nameless **CTLMode** block defines information that is to be inherited by the below (named) **CTLMode** block. The "EX_wrapper_WSP_chains" **Scan-**

Structures block used by the below CTL code is defined in Section *8.4* (see page 205).

```
<5.363>  CTLMode WP_INTEST_MBIST_MODE {
<5.364>     TestMode InternalTest;
<5.365>     DomainReferences {
<5.366>        MacroDefs ex_wrapper_macros;
<5.367>        ScanStructures EX_wrapper_WSP_chains
                              EX_wrapper_WPP_chains;
<5.368>     } // DomainReferences block
<5.369>     Internal {
<5.370>        all_functional {DataType Functional;}
<5.371>        wpp_scan {DataType Unused;
<5.372>     } // Internal block
<5.373>     PatternInformation {
<5.374>        Pattern WpIntestMbistSetupPattern {
<5.375>           Purpose EstablishMode;
<5.376>           Protocol Macro instruction_mode_setup;
<5.377>        } // Pattern block
<5.378>        Macro instruction_mode_setup {
<5.379>           Purpose ModeControl;
<5.380>           UseByPattern EstablishMode;
<5.381>        } // Macro block
<5.382>     } // PatternInformation block
<5.383>  TestModeForWrapper WP_INTEST_MBIST 0111;
<5.384>  } // CTLMode WP_INTEST_MBIST_MODE block
```

The above CTL code defines enough information to establish the WP_INTEST_MBIST test mode but does not contain information about patterns to be executed once the mode has been established.

5.5.7 Describing WP_EXTEST in CTL for the EX core

Load of the WP_EXTEST instruction is similar to the load of the WS_BYPASS instruction discussed in Section *5.5.1.1* (see page 106) except that data on the WSI terminal will be specific to the WP_EXTEST instruction. The corresponding timing diagram follows in Figure 38.

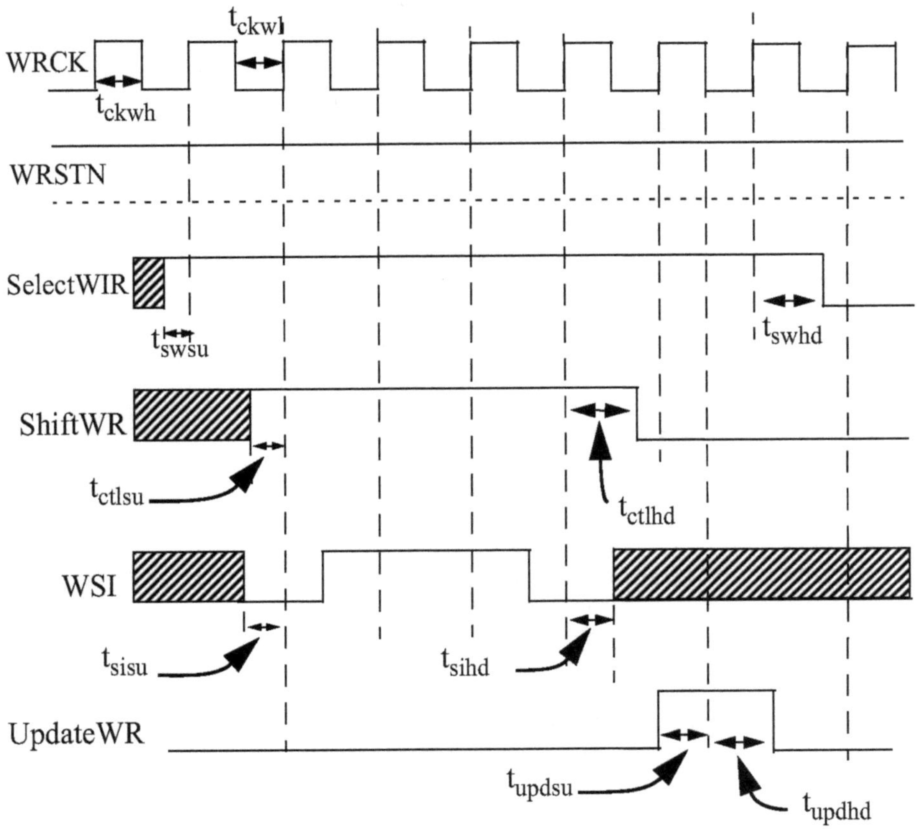

Figure 38 EX wrapper WP_EXTEST instruction load timing

5.5.7.1 Defining the CTLMode Block for the WP_EXTEST Mode

In this section the **MacroDefs** block defined in Section *5.5.1.3* (see page 109) will be reused. The pattern utilized to set up the WP_EXTEST instruction follows:

```
<5.385>   Pattern WpExtestSetupPattern {
<5.386>      P {"WSI"=0110;}
<5.387>   }
```

The following CTL code is assumed preceded by the signal group names and nameless **CTLMode** block defined in Section *4.4* (see page 68). This nameless **CTLMode** block defines information that is to be inherited by the below (named) **CTLMode** block. The "EX_wrapper_WSP_chains" **Scan-**

Structures block used by the below CTL code is defined in Section *8.4* (see page 205).

```
<5.388>  CTLMode WP_EXTEST_MODE {
<5.389>     TestMode ExternalTest;
<5.390>     DomainReferences {
<5.391>        MacroDefs ex_wrapper_macros;
<5.392>        ScanStructures EX_wrapper_WSP_chains
                              EX_wrapper_WPP_chains;
<5.393>     } // DomainReferences block
<5.394>     Internal {
<5.395>        all_functional {DataType Functional;}
<5.396>        wpp_mbist {DataType Unused;}
<5.397>           } // Internal block
<5.398>     PatternInformation {
<5.399>        Pattern WpExtestSetupPattern {
<5.400>           Purpose EstablishMode;
<5.401>           Protocol Macro instruction_mode_setup;
<5.402>        } // Pattern block
<5.403>        Macro instruction_mode_setup {
<5.404>           Purpose ModeControl;
<5.405>           UseByPattern EstablishMode;
<5.406>        } // Macro block
<5.407>     } // PatternInformation block
<5.408>     TestModeForWrapper WP_EXTEST 0110;
<5.409>  } // CTLMode WP_EXTEST_MODE block
```

The above CTL code defines enough information to establish the WP_EXTEST test mode but does not contain information about patterns to be executed once the mode has been established.

5.5.8 Describing WS_SAFE_SINGLE in CTL for the EX Core

Load of the WS_SAFE_SINGLE instruction is similar to the load of the WS_BYPASS instruction discussed in Section *5.5.1.1* (see page 106) except that data on the WSI terminal will be specific to the WS_SAFE_SINGLE instruction. The corresponding timing diagram follows in Figure 39.

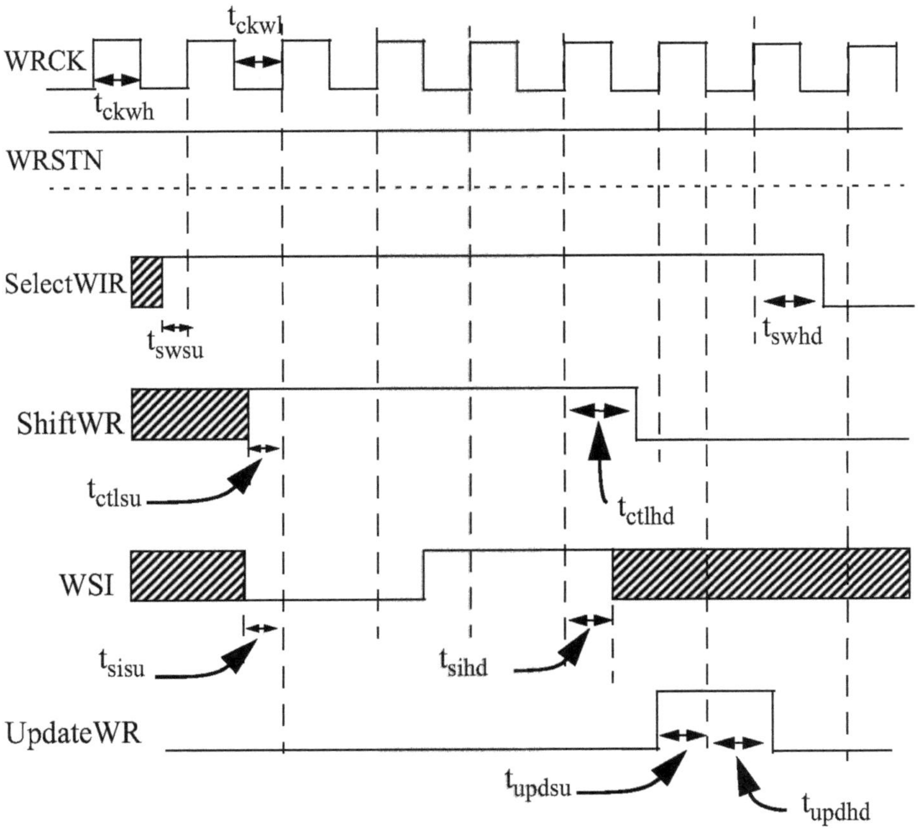

Figure 39 EX wrapper WS_SAFE_SINGLE instruction load timing

5.5.8.1 Defining the CTLMode Block for the WS_SAFE_SINGLE mode

In this section the **MacroDefs** block defined in Section *5.5.1.3* (see page 109) will be reused. The pattern utilized to set up the WS_SAFE_SINGLE instruction follows:

```
<5.410>   Pattern WsSafeSingleSetupPattern {
<5.411>       P {"WSI"=0011;}
<5.412>   }
```

The following CTL code is assumed preceded by the signal group names and nameless **CTLMode** block defined in Section *4.4* (see page 68). This nameless **CTLMode** block defines information that is to be inherited by the below (named) **CTLMode** block. The "EX_wrapper_WSP_chains" **Scan-**

Structures block used by the below CTL code is defined in Section *8.4* (see page 205).

```
CTLMode WS_SAFE_SINGLE_MODE {
    TestMode Isolate;
    DomainReferences {
        MacroDefs ex_wrapper_macros;
        ScanStructures EX_wrapper_WSP_chains;
    } // DomainReferences block
    Internal {
        all_functional {DataType Functional;}
        wpp_mbist {DataType Unused;}
        wpp_scan {DataType Unused;
    } // Internal block
    PatternInformation {
        Pattern WsSafeSingleSetupPattern {
            Purpose EstablishMode;
            Protocol Macro instruction_mode_setup;
        } // Pattern block
        Macro instruction_mode_setup {
            Purpose ModeControl;
            UseByPattern EstablishMode;
        } // Macro block
    } // PatternInformation block
    TestModeForWrapper WS_SAFE_SINGLE 0011;
    } // CTLMode WS_SAFE_SINGLE_MODE block
```

The above CTL code defines enough information to establish the WS_SAFE_SINGLE test mode but does not contain information about patterns to be executed once the mode has been established.

5.5.9 Describing WP_EXTEST_SEQ in CTL for the EX Core

Load of the WP_EXTEST_SEQ instruction is similar to the load of the WS_BYPASS instruction discussed in Section *5.5.1.1* (see page 106) except that data on the WSI terminal will be specific to the WP_EXTEST_SEQ instruction. The corresponding timing diagram follows in Figure 40.

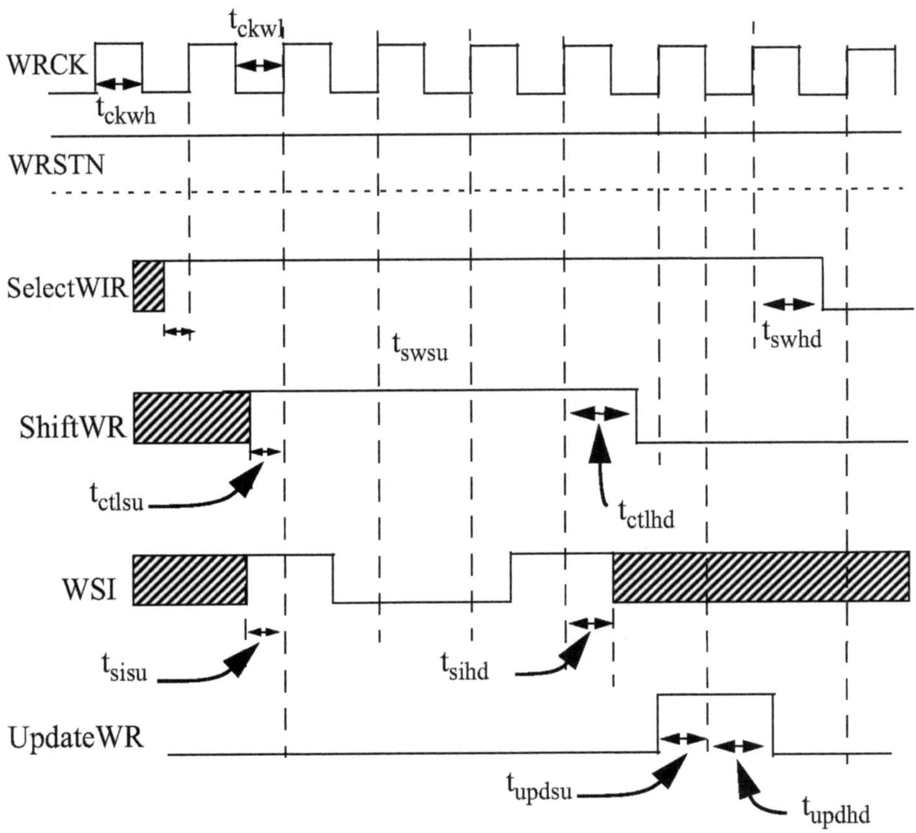

Figure 40 EX wrapper WP_EXTEST_SEQ instruction load timing

5.5.9.1 Defining the CTLMode Block for the WP_EXTEST_SEQ mode

In this section the **MacroDefs** block defined in Section *5.5.1.3* (see page 109) will be reused. The pattern utilized to set up the WP_EXTEST_SEQ instruction follows:

```
<5.436>  Pattern WpExtestSeqSetupPattern {
<5.437>      P {"WSI"=1001;}
<5.438>  }
```

The following CTL code is assumed preceded by the signal group names and nameless **CTLMode** block defined in Section *4.4* (see page 68). This nameless **CTLMode** block defines information that is to be inherited by the below (named) **CTLMode** block. The "EX_wrapper_WSP_chains" **Scan-**

Structures block used by the below CTL code is defined in Section *8.4* (see page 205).

```
CTLMode WP_EXTEST_SEQ_MODE {
   TestMode ExternalTest;
   DomainReferences {
      MacroDefs ex_wrapper_macros;
      ScanStructures EX_wrapper_WSP_chains
                        EX_wrapper_WPP_chains;
   } // DomainReferences block
   Internal {
      all_functional {DataType Functional;}
      wpp_mbist {DataType Unused;}
   } // Internal block
   PatternInformation {
      Pattern WpExtestSeqSetupPattern {
         Purpose EstablishMode;
         Protocol Macro instruction_mode_setup;
      } // Pattern block
      Macro instruction_mode_setup {
         Purpose ModeControl;
         UseByPattern EstablishMode;
      } // Macro block
   } // PatternInformation block
   TestModeForWrapper WP_EXTEST_SEQ 1001;
} // CTLMode WP_EXTEST_SEQ_MODE block
```

The above CTL code defines enough information to establish the WP_EXTEST_SEQ test mode but does not contain information about patterns to be executed once the mode has been established.

5.5.10 Describing WP_INTEST_SEQ in CTL for the EX core

Load of the WP_INTEST_SEQ instruction is similar to the load of the WS_BYPASS instruction discussed in Section *5.5.1.1* (see page 106) except that data on the WSI terminal will be specific to the WP_INTEST_SEQ instruction. The corresponding timing diagram follows in Figure 41.

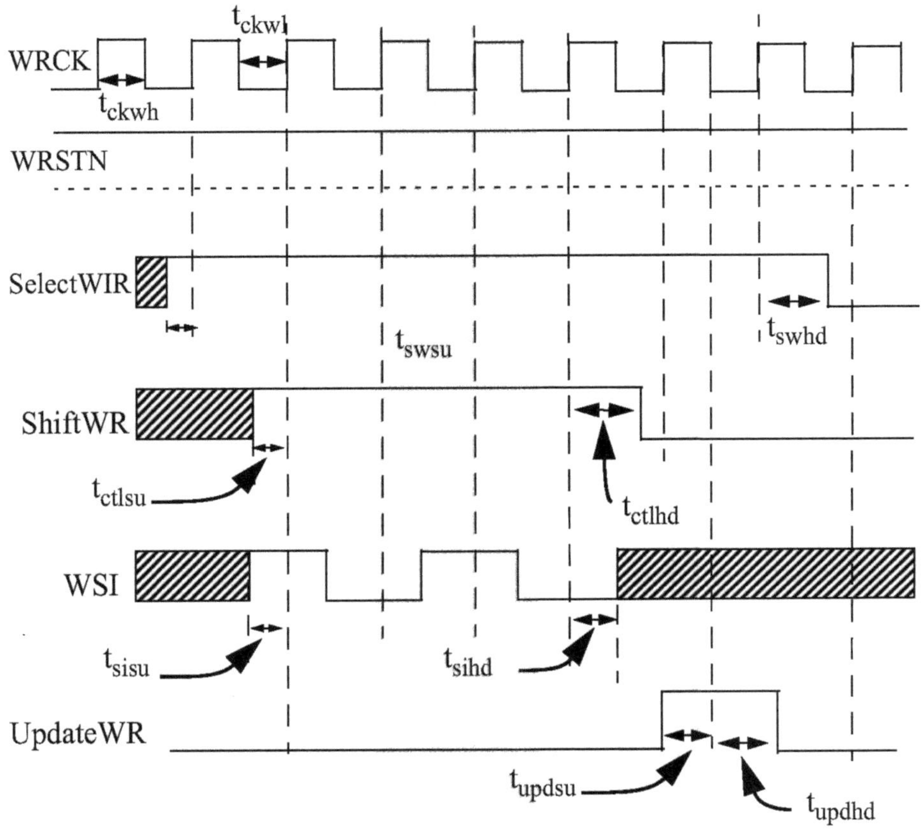

Figure 41 EX wrapper WP_INTEST_SEQ instruction load timing

5.5.10.1 Defining the CTLMode Block for the WP_INTEST_SEQ mode

In this section the **MacroDefs** block defined in Section *5.5.1.3* (see page 109) will be reused. The pattern utilized to set up the WP_INTEST_SEQ instruction follows:

```
<5.461>  Pattern WpIntestSeqSetupPattern {
<5.462>      P {"WSI"=1010;}
<5.463>  }
```

The following CTL code is assumed preceded by the signal group names and nameless **CTLMode** block defined in Section *4.4* (see page 68). This nameless **CTLMode** block defines information that is to be inherited by the below (named) **CTLMode** block. The "EX_wrapper_WSP_chains" **Scan-**

Structures block used by the below CTL code is defined in Section *8.4* (see page 205).

```
<5.464>  CTLMode WP_INTEST_SEQ_MODE {
<5.465>    TestMode InternalTest;
<5.466>    DomainReferences {
<5.467>      MacroDefs ex_wrapper_macros;
<5.468>      ScanStructures EX_wrapper_WSP_chains
                              EX_wrapper_WPP_chains;
<5.469>    } // DomainReferences block
<5.470>    Internal {
<5.471>      all_functional {DataType Functional;}
<5.472>      wpp_mbist {DataType Unused;}
<5.473>    } // Internal block
<5.474>    PatternInformation {
<5.475>      Pattern WpIntestSeqSetupPattern {
<5.476>        Purpose EstablishMode;
<5.477>        Protocol Macro instruction_mode_setup;
<5.478>      } // Pattern block
<5.479>      Macro instruction_mode_setup {
<5.480>        Purpose ModeControl;
<5.481>        UseByPattern EstablishMode;
<5.482>      } // Macro block
<5.483>    } // PatternInformation block
<5.484>    TestModeForWrapper WP_INTEST_SEQ 1010;
<5.485>    } // CTLMode WP_INTEST_SEQ_MODE block
```

The above CTL code defines enough information to establish the WP_INTEST_SEQ test mode but does not contain information about patterns to be executed once the mode has been established.

Chapter 6
Design of the WBR

The WBR enables the separation of core internal testing from external interconnect or logic testing. This wrapper register provides an isolation mechanism that allows test stimuli to propagate from wrapper test inputs to core inputs, and test response to propagate from core outputs to wrapper test outputs. WBR data inputs are normally coupled to TAMs which typically have a reduced data bandwidth compared to the data bandwidth of the embedded core being tested. As a result, the WBR serves as data bandwidth adaptation mechanism between the SOC environment and the embedded core. The presence of the WBR also enables testing of logic external to a wrapped core, as stimuli can be launched from the WBR of a given wrapper, traverse logic external to the wrapper, and be captured into the WBR of another wrapped core. This is depicted in Figure 42.

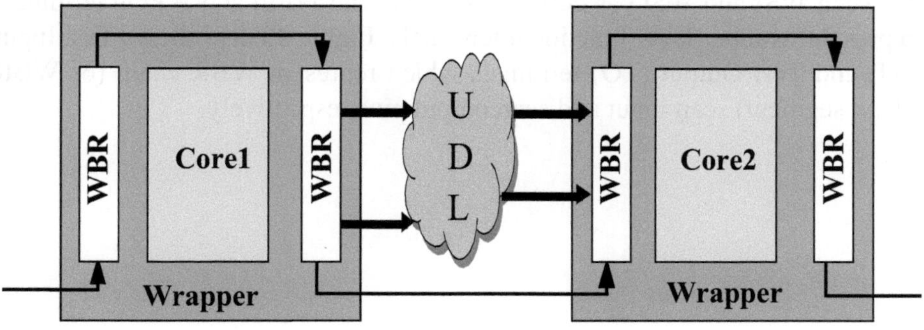

Figure 42 Illustration of external logic test via 1500 wrappers

Note: A Test Access Mechanism is a chip-level route that connects a design block to chip-level I/O.

Another role of the WBR is to shield embedded cores when SOC level activity is deemed harmful to the wrapped core or to data inside the wrapped core. Conversely, the WBR can provide protection to logic outside the wrapped core from data coming out of this core. This chapter expands on the role, design and usage of the WBR.

6.1 What is the WBR?

The WBR is a collection of 1500 compliant wrapper cells stitched into one or more scan chains inside a 1500 wrapper (definition of a 1500 compliant wrapper cell will be provided in the following sections). 1500 mandates the presence of a single-chain configuration of the WBR, but allows multiple chain implementations which are typically meant for increased test data bandwidth to and from the core being tested. As discussed in Section *1.1.1* (see page 5), test data bandwidth is a design specific requirement and therefore provision of a parallel configuration of the WBR was left optional by the standard, as was the case with provision of the WPP. In fact, provision of the WPP would typically correspond to support of a parallel configuration of the WBR and vice-versa. Requiring a single chain configuration of the WBR is a minimum condition for the support of 1500 mandatory instructions, since some instructions such as the WS_EXTEST instruction - Section *5.1.1.2* (see page 80)- use only the WSP to access the WBR. The serial configuration of the WBR is used in Plug-and-Play operations of the 1500 wrapper. Figure 43 (page 133) shows a single-chain configuration and a multiple chain configuration of the WBR. When the WPP is selected all input wrapper cells are connected between the WPI[0] and WPO[0] wrapper terminals for scan access. The output cells are connected between WPI[1], WPO[1]. During serial instructions the WPP is not enabled and all WBR cells are connected between WSI and WSO as defined by 1500. The WFIn and WFOn terminals represent wrapper level functional terminals. Figure 43 also shows Test Input (TI) and Test Output (TO) terminals which represent WBR chain (or WBR chain segment) scan input and scan output pins respectively.

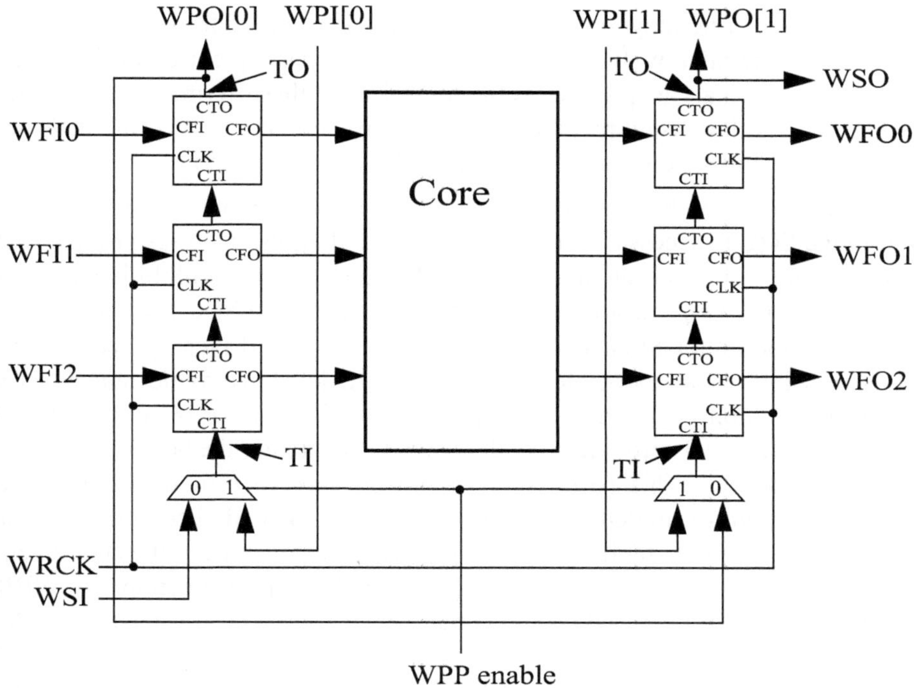

Figure 43 Example serial and parallel configuration of a WBR

6.2 Provision of WBR cells

1500 mandates provision of a WBR cell on all core-level digital terminals with the exception of clocks, asynchronous set or reset and dedicated test signals. This mandate ensures isolation, control and observation at every core-level terminal, while the exemption allows signals such as clocks and scan inputs to be directly controlled from the SOC. The result is that these exempted signals will not suffer additional propagation delays through WBR cells. However, some test strategies may require control over terminals exempted from provision of a WBR cell, and the very absence of a WBR cell makes this challenging. To address this requirement, 1500 provides support for the following wrapper structures that are less restrictive than regular WBR cells:

a. Reduced functionality cells on signals that are exempted from provision of a regular WBR cell. These cells will typically be control-only or observe-only cells provided for increased interconnect testability.

b. Harnessing logic on signals that are exempted from provision of a regular WBR cell. This is combinational logic used on WPP terminals to provide WIR control over these terminals which are otherwise directly controlled from the SOC. Harnessing cells are further discussed in Section *6.3.4* (see page 145).

Support for harnessing logic is convenient in cases where direct access from the SOC is necessary. However this type of access requires provision of an access mechanism at the SOC and this can become prohibitive when a large number of signals are involved, because of limited SOC level resources. The decision to not add a WBR cell must therefore be driven by necessity as this decision has an associated cost at the SOC level. To avoid unnecessary SOC integration challenges, 1500 permits provision of a WBR cell on all core-level terminals provided that doing so does not alter the behavior of standard instructions. This permission should be used to provide isolation on signals that are exempted from provision of a WBR cell. The permission is meant to bring predictable and often improved test coverage at the core level.

Independent of the wrapper logic (WBR cell, reduced functionality WBR cells or harnessing logic) the architecture of the 1500 wrapper must allow the WIR to remain in control of test activities inside the wrapped core. This will often imply gating non-traditional WBR cells with a control signal from the WIR.

6.3 Creation of 1500 compliant WBR cells

The goal of a 1500 compliant WBR cell is to provide isolation at the core terminal level and allow test stimuli to be applied to wrapped terminals as well as allow test response to be captured from wrapped terminals. In meeting this goal, the WBR cell must preserve access to the core in functional mode which is also referred to as the “wrapper disabled state”.

A 1500 compliant WBR cell must meet the following criteria:

a. participate in the shift operation of the WBR by containing at least one storage element on the shift path of the WBR. This ensures that test data can be applied to, and observed from the core terminal provisioned with the WBR cell.

b. contain a storage element intended for capturing data into the WBR. This ensures that test response can be loaded into the WBR from its functional path and subsequently shifted out. The storage element provisioned for capture can be a shift-path storage element but also can be an off-shift-path storage element. This will be further explained in following sections.

The above requirements essentially mandate that a 1500 WBR cell must have a shift path that is different from its functional path.

The definition of a 1500 compliant WBR cell is quite flexible compared to that of 1149.1 boundary scan cells, and allows for a variety of types of test. For example, sequential tests requiring multi-cycle captures can be supported via provision of wrapper cells with multiple storage elements in their shift path. This makes multiple cycles worth of capture data available at a given core terminal without incurring the extra delay of shifting through the WBR in-between these cycles. The flexibility in wrapper cell design makes it impossible to list every 1500 compliant WBR cell. Examples provided by the standard are only a subset of possible 1500 compliant wrapper cells.

6.3.1 WBR Cell operation

A typical WBR cell has four data terminals as shown in Figure 44 and some control terminals.

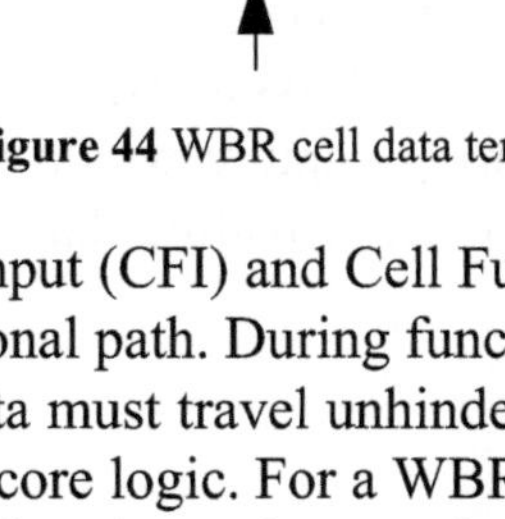

Figure 44 WBR cell data terminals

The Cell Functional Input (CFI) and Cell Functional Output (CFO) terminals connect to the functional path. During functional mode (or normal mode) of the embedded core, data must travel unhindered from CFI to CFO so as to preserve operation of the core logic. For a WBR cell provided on a core input terminal, CFO connects directly to the core input while CFI connects to the corresponding wrapper input terminal. Similarly, for a WBR cell provided on a core output terminal, CFI connects to the core output and CFO to the corresponding wrapper output terminal. This brings us to the fact that 1500 does not mandate a distinction between input wrapper cells and output wrapper cells. In fact, more often than not, identical WBR cells will be used for wrapping both core input terminals and core output terminals. However, even identical input WBR cells and output WBR cells will respond to different wrapper operations since these two types of cells serve different purposes in the wrapper. Different sets of control signals from the WIR are responsible for actuating the difference between input WBR cell and output WBR cell operation.

The Cell Test Input (CTI) and Cell Test Output (CTO) are shift path input and shift path output terminals to the WBR cell. For WBR cells containing a single storage element, CTI and CTO connect to the storage element's scan input and scan output pins respectively. For WBR cells with multiple storage elements in the shift path, CTI connects to the scan input pin of the first storage element on the shift path while CTO connects to the scan output pin of the last storage element on the shift path. At the WBR level, multiple WBR cells are concatenated into one or more shift paths through the CTI and CTO cell terminals.

While 1500 defines the name of data terminals (CFI, CFO, CTI and CTO) required to operate WBR cells, creation and naming of the control signals required at the cell level is left to the user. These control signals are expected to be sourced from the WIR and enable the shift, capture, update and transfer operations at the WBR cell level.

6.3.1.1 Providing support for the shift operation

The Shift event allows data to move from the WBR's TI to the WBR's TO one storage element at a time. This is allowed by the shift path storage elements that exist inside each WBR cells and participate in the Shift operation. The number of such storage elements present in a WBR cell will depend on the types of test intended for the core being wrapped. The main consideration is that test data required at a given core terminal is contained solely in the WBR cell provided for this terminal. As a result, sequential tests having multiple cycles worth of data during capture would require multiple shift path storage elements in all WBR cells providing data for this test, see Figure 45 (page 137).

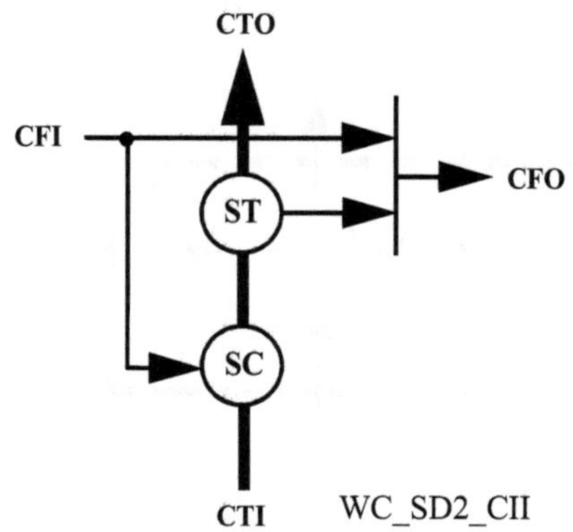

Figure 45 WBR cell with multiple storage elements on shift path

6.3.1.2 Providing support for the capture operation

At the WBR cell level, the capture operation allows data to enter the WBR cell either from CFI, see Figure 46, or from CFO, see Figure 47. Capturing from CFO mainly provides observability to the cells path going through CFO. The location of captured data is restricted to the first storage element of the WBR's shift path, its last storage element as show in Figure 48, or the optional update storage element if provided, see Figure 49 (page 139).

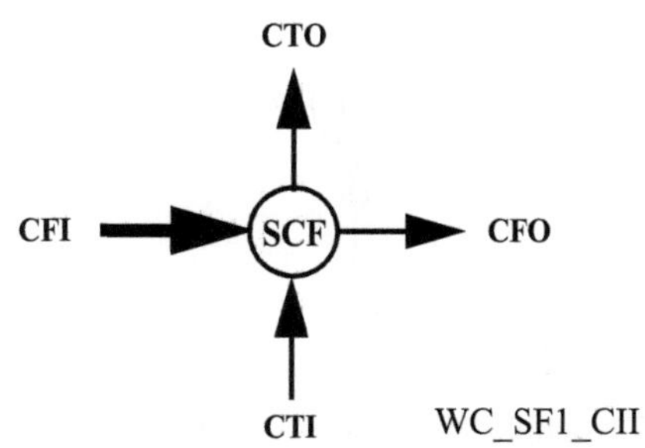

Figure 46 Example WBR cell with capture from CFI

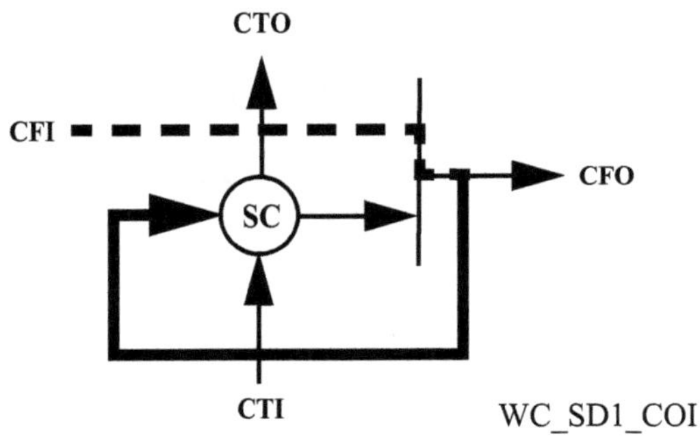

Figure 47 Example WBR cell with capture from CFO

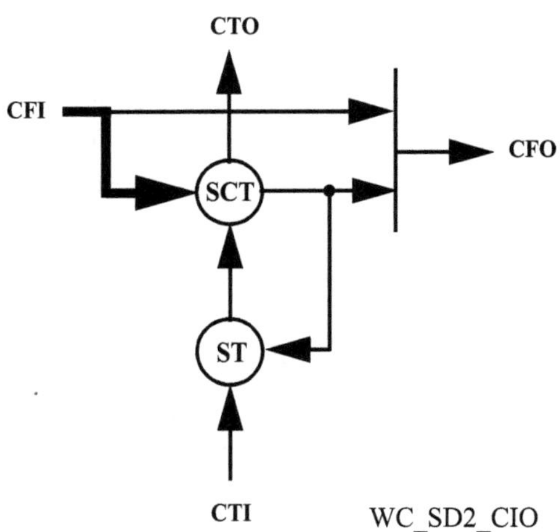

Figure 48 Example WBR cell with capture in last storage element

6.3.1.3 Providing support for the update operation

An optional update register may be provided in WBR cells. This register is an off-shift-path storage element intended for an update operation, and that is allowed to capture data from CFI. The 1500 support for update registers is similar to that of 1149.1. Additionally, the update storage element can be a shared storage element (a functional storage element reused during the implementation of the WBR cell). An example of WBR cell that uses a functional storage element for the update operation is shown in Figure 49. In addition, the capture operation is supported by the same element. The movement of captured data from an off-shift-path storage element into a shift path storage element is referred to as the transfer operation.

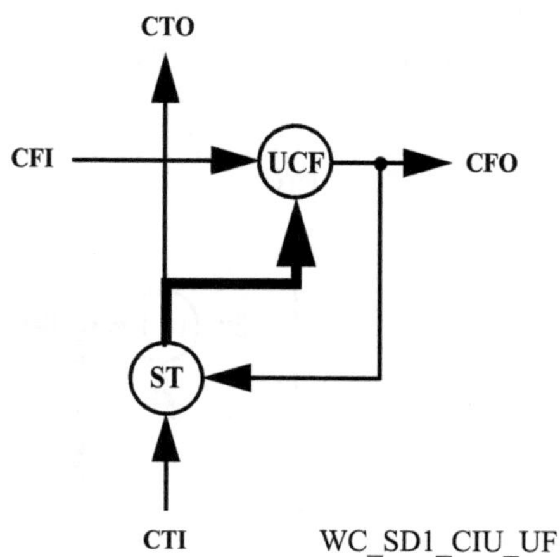

Figure 49 Example WBR cell with an update stage

6.3.1.4 Providing support for the transfer operation

Support of the Transfer operation is optional at the WBR cell level. This operation is used in WBR cells such as that shown in Figure 50 to move data through the cell. There are two possible data movements that qualify as transfer, both are shown in Figure 50:

- **a.** the case where an update register is provided in the WBR cell for the purpose of capturing data into the cell. In this case, the location of the capture data is an off-shift-path storage element and this data must be brought to the shift-path storage element closest to CTI so that it can be shifted out of the WBR cell and subsequently be observed. Similarly, when the capture site is the shift-path storage element closest to CTO, the capture data needs to be moved into the shift-path storage element closest to CTI so as to be properly shifted out. The action of getting capture data, from the two capture sites cited above, into the shift-path storage element closest to CTI is referred to as transfer.
- **b.** the case where a WBR cell is designed with multiple storage elements on the shift path. In this scenario, the shift operation local

to the WBR cell and allowing data to move from CTI to CTO is referred to as transfer

Note: Within a wrapper or SOC there may be WBR cells that support the transfer operation and others that do not. For this reason, WBR cells without support for transfer must hold data during Transfer in order to preserve the integrity of the test at the wrapper level.

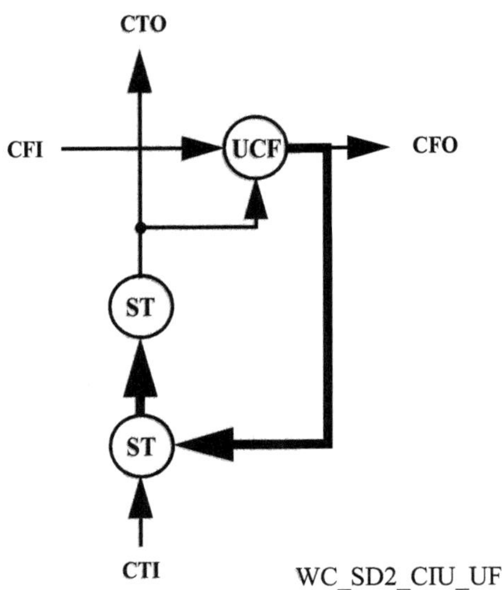

Figure 50 Example WBR cell supporting transfer

6.3.2 Dedicated WBR Cell implementation

A dedicated WBR cell implementation does not make use of any functional storage element but uses storage elements dedicated to 1500 operations. The resulting cell can be used for isolation of all types of functional core terminals.

In Figure 51 a core with two categories of terminals (registered and unregistered) is depicted. A registered input port directly connect to a register without passing through a combinational logic tree. A registered output port is directly sourced from a functional sequential element. A non-registered port will be referred to as 'Unregistered'.

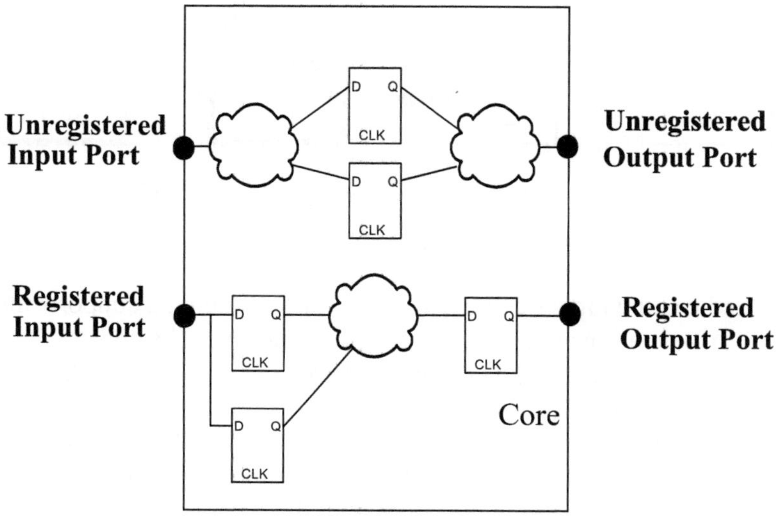

Figure 51 Registered and unregistered functional core terminals

Unregistered ports are expected to be isolated by dedicated WBR cells in order to guarantee controllability and observability during test. In contrast, the storage element associated with registered ports can be reused to provide the control and observe capabilities required for isolation. This means that providing a WBR cell for registered ports may only involve adding the logic required to transform the functional register into a compliant WBR cell. For example, adding the mandatory hold behavior at the WBR cell level would be required for a shared WBR cell.

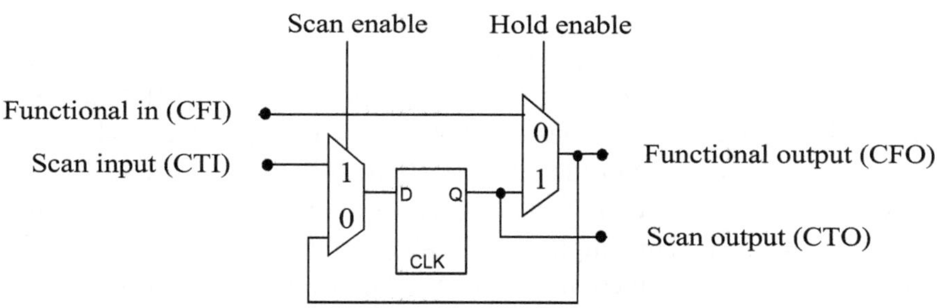

Figure 52 Dedicated wrapper cell example

Figure 52 shows a dedicated WBR cell implementation. This cell can be used for both core input and output terminals.

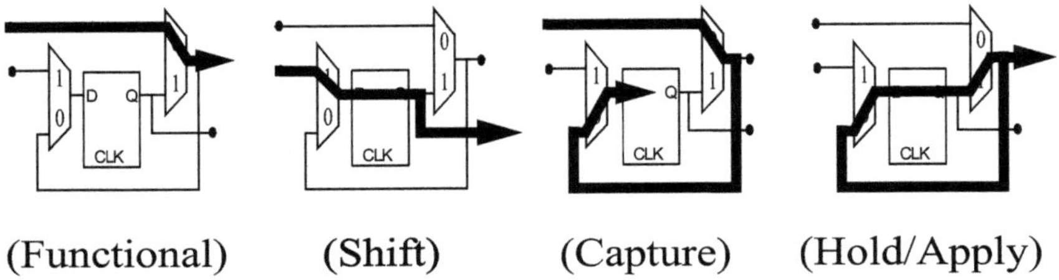

Figure 53 Example of dedicated wrapper cell operations

The dedicated wrapper cell example shown in Figure 52 supports the minimum set of operations required for a 1500 WBR cell. The control values for the cell operations are listed in Table 9 and corresponding active datapaths are highlighted in Figure 53.

For testability purposes it is important to note that all nets involved in the design of the WBR cell shown in Figure 52 are exercised by the various Shift, Capture and Hold datapaths. This means that the complete structure including the functional path is tested by invocation of the shift, capture and hold/apply operations during test.

Table 9 Example of dedicated wrapper cell operation control

WBR cell Operation	*Scan enable*	*Hold enable*
Functional	-	0
Shift	1	-
Capture	0	0
Hold / Apply	0	1

Reuse of functional registers in the design of wrapper cells provides area reduction benefits in addition to preservation of timing-critical paths. However, this carries design considerations that must be addressed in order to achieve compliance with the 1500 standard. An example of such a compliance-related consideration is that during serial instructions the WBR must use the WRCK clock. This implies that additional circuitry must be added to the wrapper so that all functional elements that are shared for test isolation are able to respond to WRCK. This logic may challenge timing analysis and timing closure. On the other hand, a dedicated wrapper cell adds more delay on the functional data path and takes more area than a shared wrapper cell.

The choice between shared and dedicated wrapper cells should result from a cost analysis aimed at determining the least costly of the two implementations. This analysis comes in addition to a wrapper type selection that is based on the type of test required for the core. Also, it should be noted that the presence of registered ports does not necessarily equate to the use of shared wrapper cells. Figure 54 shows how dedicated wrapper cells can be added to isolate registered ports. This results in a straightforward WBR implementation with a unique clock timing for all WBR cells, but incurs an additional mux delay on the functional path going through the WBR cell. Figure 54 also depicts that provision of a 1500 wrapper cell is on a per-terminal basis independent of the number of data paths to, or from, each terminal.

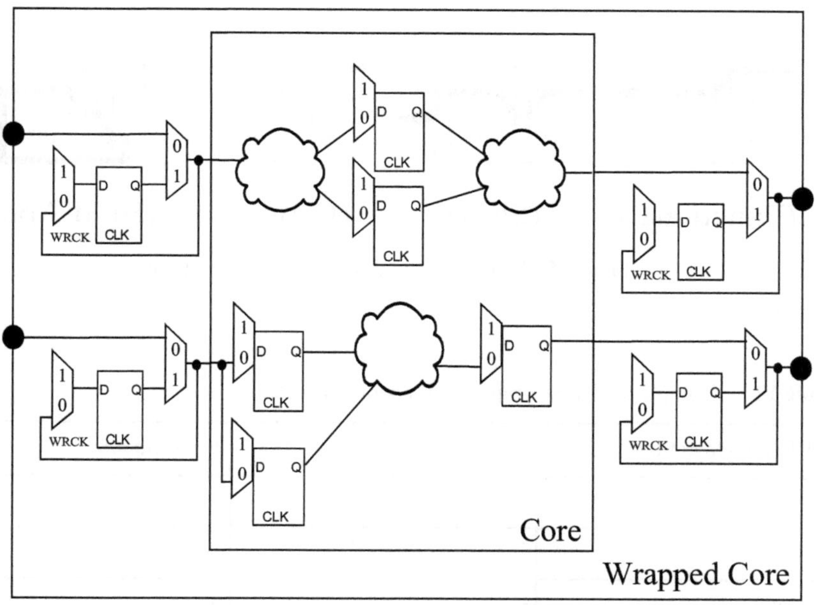

Figure 54 Example dedicated wrapper cell usage on registered ports

6.3.3 Shared WBR Cell implementation

Figure 55 shows a shared WBR cell implementation. This implementation can be used for registered core input and output terminals. Whether the cell acts as input or output isolation is naturally defined by the functional paths of the shared storage element. Of course, the proper control values need to be applied during WBR Inward Facing and Outward Facing operations. These modes are defined in Section *6.5.1* (see page 152).

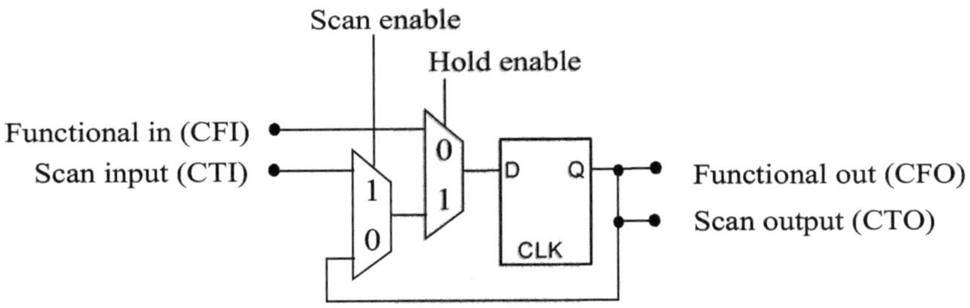

Figure 55 Example of shared wrapper cell

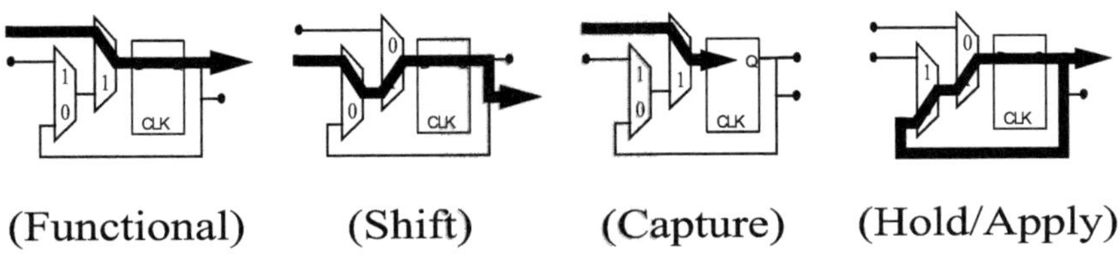

Figure 56 Example shared wrapper cell operations

Table 10 Example of shared wrapper cell operation control

WBR cell Operation	*Scan enable*	*Hold enable*
Function	-	0
Shift	1	1
Capture	-	0
Hold / Apply	0	1

The shared wrapper cell example shown in Figure 55 supports the minimum set of operations required for a WBR cell. The control values required for operation of the cell are listed in Table 10 and corresponding active datapaths can be found in Figure 56.

It is important to observe that similar to the dedicated wrapper cell example, the entire structure of the WBR cell is fully testable by a combination of Inward Facing and Outward Facing tests. It should also be noted that the functional clock source is overruled by WRCK during serial 1500 instructions. As a result, part of the functional path might be left untested depending on the implementation of the WBR cell.

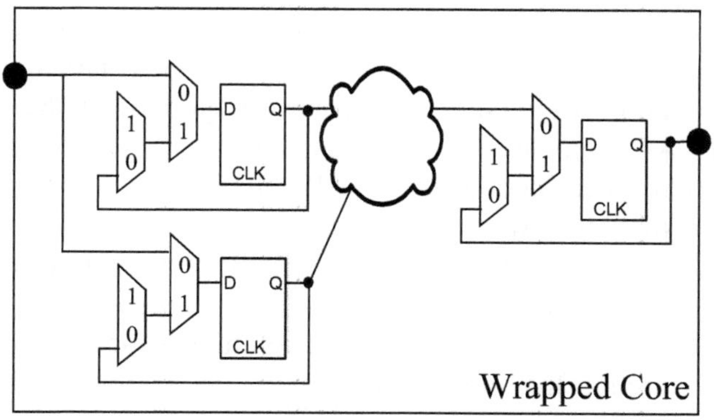

Figure 57 Example shared wrapper cell usage

There is a subtle difference between the usage of shared wrapper cells and that of dedicated wrapper cells when an input terminal feeds into multiple sequential elements. In the case shown in Figure 57, these elements need to be part of the WBR in order to reach full test coverage of the functional core paths during external testing. In the corresponding dedicated wrapper cell usage, only one WBR cell is added as shown in Figure 54 (page 143).

6.3.4 Harness Cells and Reduced Functionality Cells

Core inputs and outputs that are used only during test can be wrapped by harness cells or reduced functionality WBR cells.

A harness cell does not make use of any functional logic and/or functional storage element. In Figure 58 two 'pass-through' harness cells are shown.

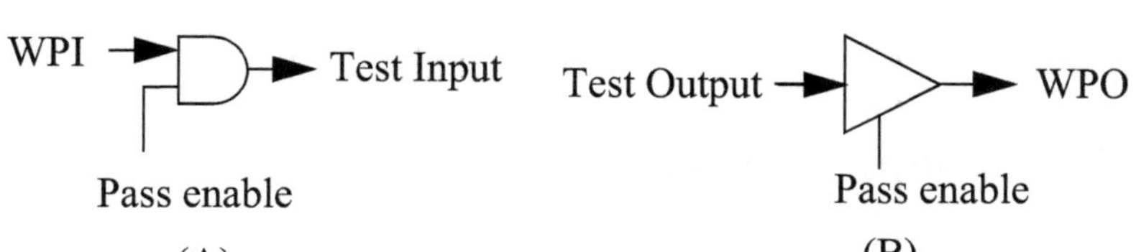

Figure 58 Pass-Through Harness Cells

The input cell (A) connects the core test input to the wrapper terminal WPI when the "Pass enable" signal is high. Under the same condition the output cell (B) connects core test output directly to wrapper output WPO. The core test input is driven to a low value and the wrapper terminal WPO is tri-state while the "Pass enable" signal is low. This is typically the case when the core operates in mission mode.

Reduced functionality WBR cells are typically observe-only or control-only cells. These cells do not qualify as full WBR cells typically because they do not meet both of the (shift and capture) requirements that apply to 1500 WBR cells. An example of such cell is shown in Figure 59 which depicts a control-only reduced functionality WBR cell. Reduced functionality cells are useful for providing test coverage through the WBR. For this reason, it is required that these cells participate in the Shift operation of the WBR so that test stimuli can be shifted into them or so that test response can be shifted out of these cells. In addition, a control-only cell must support application of data at its CFO (through the apply event) and an observe-only cell must support the capture event.

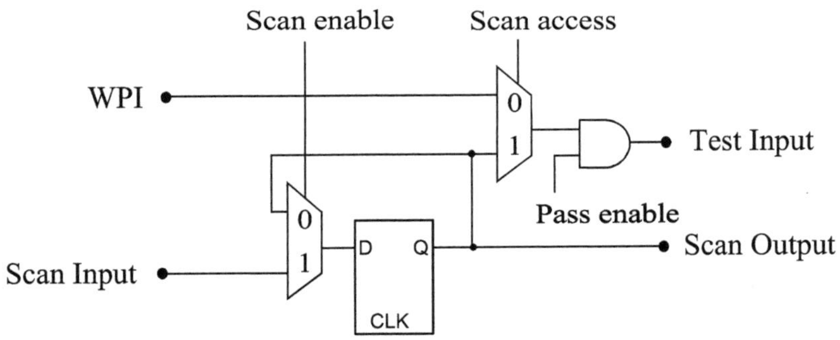

Figure 59 WC_SD1_CN_G cell

The cell shown in Figure 59 has a scan access mechanism that is controlled by the "Scan access" terminal. This terminal is driven by the WIR circuitry.

6.4 Describing wrapper cells in 1500

In general, 1500 describes only mandatory behavior of WBR cells and additional cell capabilities may be added by the user while preserving compliance to the standard. As a result, the universe of 1500 compliant WBR cells is not limited by the standard and multiple implementations can be created for a given WBR cell type. This created a need to express WBR cell types using a standard naming convention since cell types selected by a core provider will ultimately need to be communicated to the core user. This naming convention also enables ease of automation because the resulting WBR cell names can be parsed by automation tools and as a result the characteristics of WBR cells can be identified by these tools.

Figure 60 shows implementation of an example WBR cell and the corresponding 1500 name. An experienced reader will recognize this cell as the BC_1 cell defined in 1149.1. It should not be a surprise that this cell is also a compliant 1500 WBR cell.

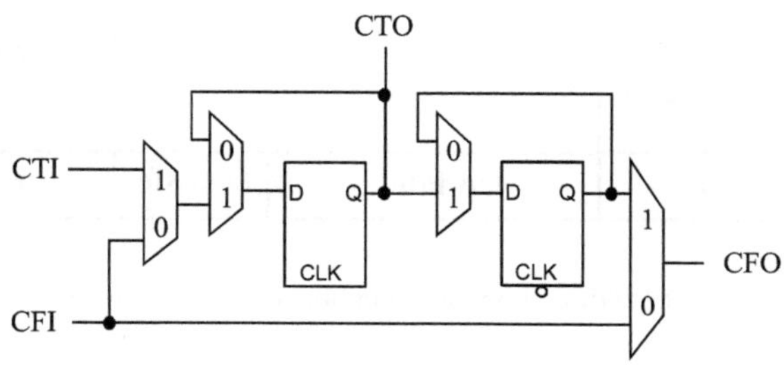

Figure 60 Example implementation WBR cell: WC_SD1_CII_UD

6.4.1 Describing WBR cells

In the process of understanding the 1500 WBR cell naming convention, it is important to note that the 1500 working group made explicit efforts to remain implementation independent by focusing exclusively on description of cell behavior. To this end, an implementation-independent representation of WBR cells was adopted. This representation is referred to as "bubble diagram".

6.4.1.1 Understanding bubble diagrams

Figure 61 shows building blocks for bubble diagrams. A circle represents a storage element and a line with an arrow head represents a data path. The storage element carries one or more letters indicating the WBR events that are supported by this storage element. The hold operation is implicitly supported by all WBR storage elements and is therefore not represented.

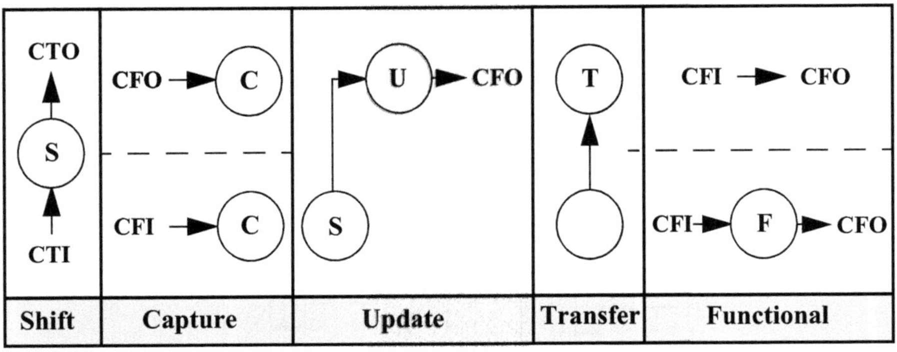

Figure 61 WBR cell operation bubble diagram representations

During a Shift operation data flows from CTI to CTO and can go through multiple storage elements. During a Capture operation data can be captured from either the CFO or CFI terminal into a storage element.

Update data is transported from the shift path storage element closest to CTO and applied to CFO. During a transfer operation data is always taken over from another storage element and not from a cell terminal. Functional data floats from CFI towards CFO. A functional storage element might exist in-between CFI and CFO. The apply event is a virtual event composed of the other events and therefore not represented explicitly in a bubble diagram.

Bubble diagrams are created by merging the building blocks depicted in Figure 61. A vertical line is used to represent a data flow decision point such as a mux.

Figure 62 gives the bubble diagram of a WC_SD1_CII_UD example cell comprising two dedicated storage elements. The first storage element supports both shift and capture from CFI terminal. The bubble for this storage element is created by merging the shift and capture building blocks of Figure 61. The second storage element supports only the update operation. Adding the functional building block (path from CFI to CFO) to the bubble diagram completes this diagram of Figure 62. Since both the functional building block and the update building block assign data to the CFO terminal, a dataflow decision point must be added. This is shown in Figure 62. Additional examples are given in the 1500 standard document.

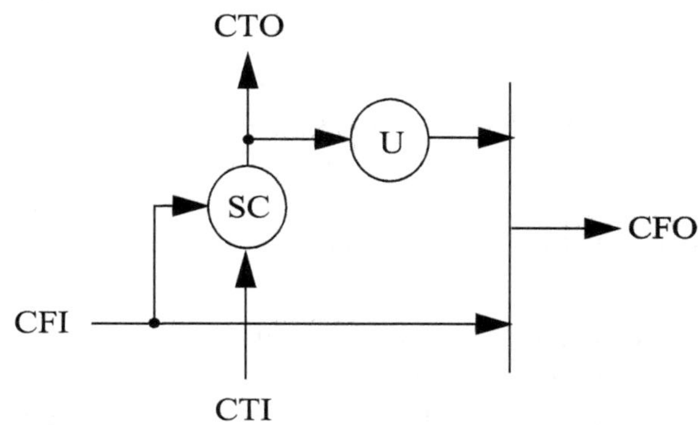

Figure 62 Bubble diagram for WBR cell: WC_SD1_CII_UD

6.4.1.2 WBR cell naming convention

The 1500 WBR cell naming convention defines a specific sequence of characters as follows:

1. The first character field, WC (for Wrapper Cell) is mandatory for WBR cells and signifies that a wrapper cell is being defined.
2. The second character field is the mandatory shift field. This field starts with "_S" followed by a letter that indicates whether the cell contains storage elements that are Dedicated to test (_SD) or shared with Functional mode (_SF). This field ends with a number that specifies the number of shift path storage elements that exist in the WBR cell. The regular expression corresponding to this field is "/ _S[DF]\d+/" which matches "_SD1" in the example shown in Figure 62.
3. The third character field is the mandatory capture field. This field starts with "_C" and describes data captured into the cell. Two additional letters follow in this field and describe the source terminal of the captured data as well as the location of the storage element serving as capture site.
   The source terminal can be "I" for functional Input i.e. CFI, "O" for functional Output i.e. CFO, or B for Both CFI and CFO. Use of "B" represents cells that can selectively capture from CFI or CFO.
   The capture site can be "I" for the shift path input cell (cell connected to CTI), or "O" for the shift path output cell (cell connected to CTO), or "U" for the optional update storage element. Note that the latter implies a cell that supports the transfer event to move the capture data

into the shift path. A reduced functionality WBR cell that is meant only for controlling will have neither a capture source nor a capture site. This type of cells is represented by "N" (No capture) in place of the capture source and capture site letters. The resulting capture field of the WBR cell name for this type of cell will therefore be "_CN". This essentially represents a control-only cell. The overall regular expression corresponding to the capture field is "/_C([IOB][IOU])|N/ ", which matches "_CII" in the example shown in Figure 62 (page 149).

4. The fourth character field is the optional update field. This field starts with "_U" and is followed by D (for an update storage element Dedicated to test) or F (for a Functional update storage element). The regular expression corresponding to this field is "/(_U[DF])?/", which matches "_UD" in the example shown in Figure 62.
5. The fifth character field is the optional observe-only field. This field is reserved for observe-only cells and has a regular expression of "(_O)?".
6. The last field is the optional safe data field. This field starts with "_G" (for Guarding data) and can optionally be followed by the guarding or safe value output by the cell. The regular expression corresponding to this field is "/(_G[01]?)?/".

Note: None of the above fields can be repeated in a given WBR cell name.

6.4.1.3 Harnessing logic naming convention

The 1500 wrapper harnessing logic naming convention defines a specific sequence of characters as follows:

1. The first character field WH (for Wrapper Harness) is mandatory in the description of harnessing logic.
2. The second character field is mandatory and expresses the combinational nature of harnessing logic. This field starts with "_C" and can be optionally followed by "I" (for inversion) in cases where output data from the cell is inverted with respect to its input data. The regular expression corresponding to this field is "/_CI?/".
3. The third character field is the mandatory capture field. This field is expected to be "_CN" (No capture) to signify that the harnessing logic does not capture data. The regular expression corresponding to this field is "/_CN/"

4. The last field is the optional safe data field. This field starts with "_G" (for Guarding data) and can optionally be followed by the guarding or safe data output by the cell. The regular expression corresponding to this field is "/(_G[01]?)?/"

Note: None of the above fields can be repeated in a given WBR cell name.

6.5 WBR Operation Events

The 1500 WBR differs from the 1149.1 boundary scan register in that the boundary scan standard prescribes a finite state-machine that forces mutually exclusive operations in predefined sequences. The 1500 wrapper does not come with a state-machine for control but relies solely upon WSC and WPC terminals to initiate various operations inside the wrapper. The behavior of the WBR is expressed in events that are activated by WRCK when enabled by the other WSC or WPC terminals. Pre-defined WBR events are Shift, Capture, Update, Transfer and Apply. The Apply event is a 'virtual' event in the sense that it refers to any optional operations that makes WBR data ready for usage following Shift, Capture, Update and Transfer.

The flexibility in WBR cell design implies that different types of WBR cells may exist in a given WBR chain, with some cells supporting events that others do not support. To enable the co-existence of these cells, 1500 mandates that while the WBR is performing an event that is not supported by a particular WBR cell, this cell is expected to execute a "hold" operation. For instance, in a WBR that performs the update operation, all WBR cells without an update stage must hold their content while other cells with an update stage are loading their update register. Similarly, in the absence of an active wrapper event, the entire WBR is expected to hold its state. Absence of an active wrapper event can be characterized by absence of a WRCK or functional clock event or by absence of an active value on a WSC or WPC terminal. Also, in order to preserve integrity of test data at the WBR level, 1500 mandates that the WBR Shift event not occur simultaneously with the WBR cell Capture, Update or Transfer events. This means that no other events can occur during Shift. However, during Capture, Update or Transfer other events such as hold can occur.

When the WIR is selected by assertion of the SelectWIR terminal, the WBR is allowed to change state. However, in some cases the WBR is required to hold its content for usage by a subsequent instruction. For instance, when update stages of a WBR are pre-loaded by an instruction, these stages must hold their content while the WIR is loading the following instruction that will make use of the pre-loaded values.

With the exception of the Apply and Shift events, all WBR events are defined at the WBR cell level as shown in section 6.3.1.

6.5.1 WBR Operation Modes

The WBR has three mandatory modes; Normal mode, Inward Facing (IF) mode, and Outward Facing (OF) mode. In Normal mode the wrapper is transparent. This mode is meant for functional operation, i.e., when the core is not tested. IF mode allows test of the wrapped core through application of test data to core input terminals and capture of test responses from core output terminals as shown in Figure 63. The OF mode allows test of logic external to the wrapped core through application of stimuli from wrapper output terminals and capture of responses into wrapper input terminals as shown in Figure 64.

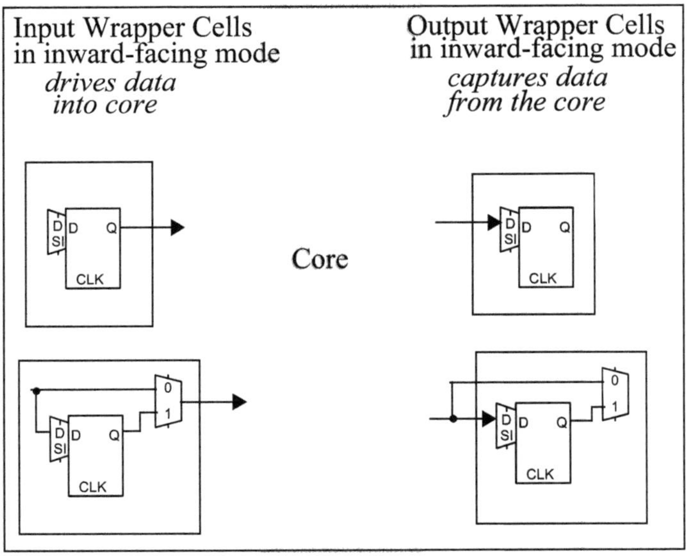

Figure 63 WBR cells in Inward Facing mode

During Inward Facing mode, an input WBR cell must support:

- **a.** the shift operation under control of the WIR circuitry
- **b.** the ability to hold data loaded into the cell
- **c.** store data meant to be applied at the CFO terminal.

In the same mode, an output WBR cell must support:

- **a.** the shift operation under control of the WIR circuitry
- **b.** the ability to hold data loaded in to the cell
- **c.** the capture operation meant to store response data from CFI into the WBR cell.

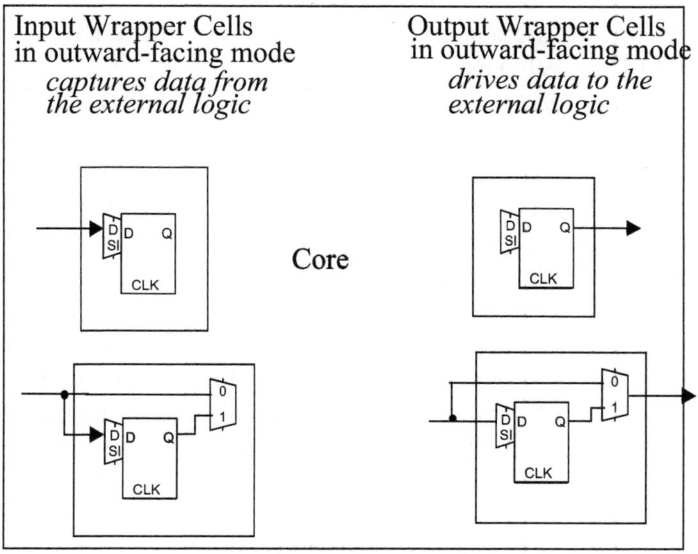

Figure 64 WBR cells in Outward Facing mode

During Outward Facing mode an input WBR cell must support:

- **a.** the shift operation under control of the WIR circuitry
- **b.** the ability to hold data loaded into the cell
- **c.** Capture data from CFI into the WBR cell

In the same test mode an output WBR cell must support:

- **a.** the shift operation under control of the WIR circuitry
- **b.** the ability to hold data loaded into the cell
- **c.** the ability to apply at CFO, data stored into the WBR cell.

In addition to the Normal, IF and OF modes described above, the WBR can implement other user defined modes for usage during private user instructions. An additional mode described by 1500 is the Non-hazardous mode. This mode forces a configuration in which the wrapped core does not disturb data present in logic external to the 1500 wrapper. The wrapper outputs 'safe' data in this mode and the embedded core should draw as little current as possible. The safe data is meant to prevent undesirable occurrences like bus contention and excessive noise creation.

6.5.2 Parallel configuration of the WBR

Parallel configuration of the WBR is enabled by the WIR in response to a parallel or hybrid instruction. During these instructions, WBR segments can still use WRCK but are otherwise fully under control of WPC terminals. All parallel instructions are also expected to configure the WBR so that it uses data from WPI and WPO exclusively.

Implementing wrapper parallel ports to increase data throughput requires knowledge of the environment in which the core will be used. In the case of embedded cores that are fully tested by BIST structures the need for parallel access is limited to test time improvement of the interconnect logic test. For embedded logic cores that need scan access for both internal testing and external testing, the parallel configuration of the WBR in combination with the configuration of internal scan chains is more challenging. In this case, the configuration allowing the smallest test time can be created only when the IC pins available for test access are known. Since this is unlikely in a core re-use environment, 1500 allows multiple parallel configurations that can be enabled by multiple parallel or hybrid instructions. This allows implementation of different width interfaces in one wrapper and provides more flexibility to the IC integrator.

Figure 65 (page 155) shows an example wrapper with a WBR chain that is configurable. The configuration is done by two sets of multiplexers.

The first set is controlled by the *Extest enable* signal. These multiplexers are used to concatenate internal scan chain segments with wrapper chain segments. The IF mode test requires access to core registers and wrapper registers. The OF mode test requires access to wrapper registers. For this reason, the internal scan chain elements are bypassed when the *Extest enable* is asserted.

The second set of multiplexers is controlled by the *WPP enable* signal. These multiplexers are used to concatenate the WBR chain or allow it to be divided into multiple segments.

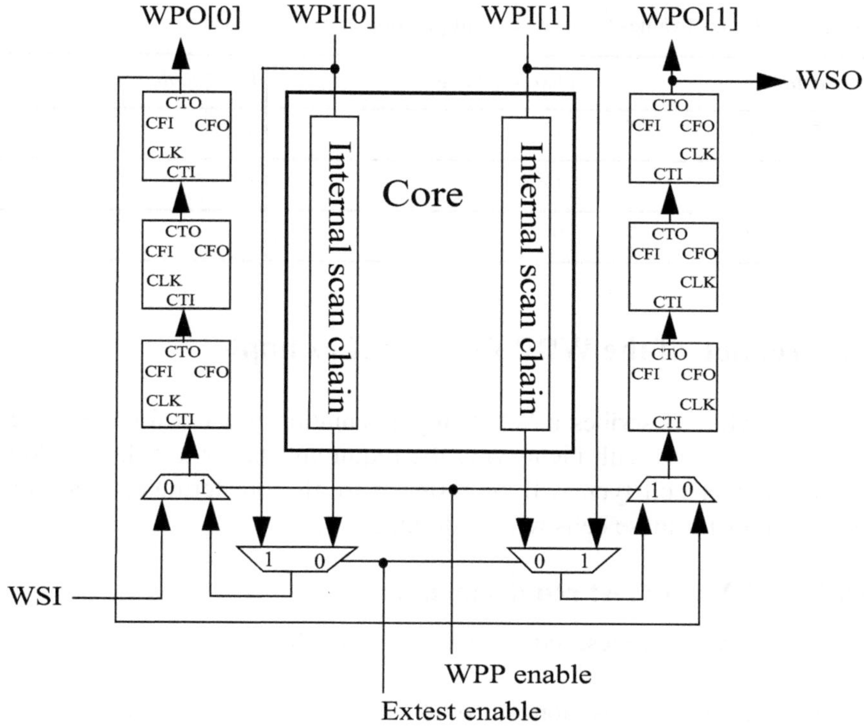

Figure 65 Example parallel configuration of core chains and WBR

Table 11 shows instructions and accompanying control signal values for the wrapper that is shown in Figure 65. During the mandatory WS_EXTEST instruction, all WBR cells are connected in a single chain between WSI-WSO for serial access. The wrapper also supports the WP_EXTEST instruction which makes use of a WPP that consists of two scan input terminals (WPI) and two scan output terminals (WPO). During the WP_EXTEST instruction all WBR cells are part of a wrapper chain segment that can be accessed either via WPI[0]-WPO[0] or via WPI[1]-WPO[1] terminals. Core registers are bypassed during this instruction. These terminals also provides high data bandwidth in IF mode scan tests in order to limit test execution time. For this purpose, during WP_INTEST, all WBR cells and internal scan chains are part of chain segments that make use of the full WPP bandwidth in a manner similar to the WP_EXTEST configuration.

Table 11 Control values for wrapper configuration in parallel mode

Instruction	*Extest enable*	*WPP enable*
WS_EXTEST	1	0
WP_EXTEST	1	1
WP_INTEST	0	1

6.6 Creation of the WBR for the EX example

This section describes a WBR implementation process for the EX core. This process starts with identifying the requirements for test. In a following step a wrapper cell type will be assigned to all core terminals and control requirements for these cells will be defined.

6.6.1 EX core test requirements

The following are test requirements for the EX core:

- Memory test is done through the built-in MBIST engine. During BIST execution, this engine needs direct access from pins to core terminals. The memories are isolated from surrounding logic during BIST runs. The WBR is in transparent mode.
- Scan tests for the core internal logic will be created by standard ATPG and executed using a standard scan protocol. This means a single 'scan enable' source for all chains (core internal and wrapper) will be used to select between scan shift cycles and scan capture cycles.
- Scan tests for the core external logic will be executed using a standard scan protocol or a finite state machine protocol. e.g. TAP controller.
- The functional terminals of the EX core are timing critical. As a result, terminals that are registered will be isolated using a shared WBR cell implementation to minimize the delay introduced by test logic.
- All core internal logic must be tested via scan. This includes both functional logic as well as dedicated test logic e.g. the BIST engine.

On the EX core, all of the functional inputs and outputs are registered, except for CLK, RESET and READY. RESET, due to the fact that it is asynchronous, and CLK do not require wrapper cells. However, for RESET an observe-only cell will be implemented in order to increase test coverage as described in Section *6.6.4* (see page 162). A dedicated WBR cell will be attached to the READY terminal in order to control and observe the unregistered logic connected to that port. All other functional ports, except for the BC port, will each be assigned to a shared wrapper cell. The BC port controls a bus and as a result needs a special WBR cell to enable a standard scan protocol during scan test. Isolation of this terminal is discussed in Section *6.6.4* (see page 162).

In addition to functional terminals, the EX core has terminals dedicated to test. These types of terminals do not require wrapper cells. To enable testing of the MBIST logic, dedicated wrapper cells will be added to the MBIST control terminals. These wrapper cells will enable control and observe during scan test and will be transparent during MBIST execution. The scan-in and scan-out terminals will not be provisioned with wrapper cells. Static test mode control signals will be directly controlled from the WIR as required by the standard. Control of the scan enable terminal is discussed in Section *8.3.3.4* (see page 209) together with control of the static test mode signals.

6.6.2 EX core WBR cell design

To begin the wrapper design process, the first step to be taken is to decide the type of wrapper cell desired for each functional port and how these cells will be implemented.

6.6.2.1 Dedicated WBR cell assignment

The following EX core terminals will be provisioned with dedicated WBR cells: READY, MBISTDLOG, MBISTRUN, MBISTDLOGOUT, MBISTDONE, MBISTFAIL. The corresponding cell type will be the WC_SD1_COI cell introduced in section 6.3.2. This cell is shown in Figure 66 (page 157) and the accompanying bubble diagram is depicted in Figure 67 (page 159).

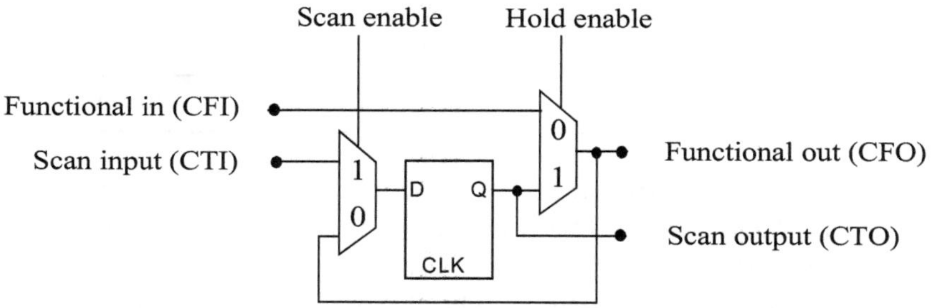

Figure 66 Implementation WC_SD1_COI

Table 12 shows control signal values for dedicated WBR cells for all instructions that will be implemented in the EX core wrapper. This table illustrates that input cells receive controls that differ from controls received on output cells. This will be the only difference between input and output cells as both types of cell will share the same design. WSC terminals will control the wrapper and core chain operation during serial instructions (WS_EXTEST, WS_INTEST) requiring scan access through the WBR. During parallel instructions, the WPC will provide control to the WBR. For the EX core wrapper, a single WPC terminal called Wrapper Parallel Scan Enable (WPSE) will be implemented. A single WPC terminal is sufficient because the various tests implemented for the EX core will use a regular scan protocol for which only a single control terminal acting as scan enable is required. The 1500 standard allows the use of multiple scan enables, for instance, to keep scan control for the wrapper separate from that of internal chains.

Table 12 Control values for WC_SD1_COI

	Input cell		*Output cell*	
Instruction	*Scan enable*	*Hold enable*	*Scan enable*	*Hold enable*
WS_EXTEST	*ShiftWR*	*~CaptureWR*	*ShiftWR*	1
WS_INTEST	*ShiftWR*	1	*ShiftWR*	*~CaptureWR*
WP_EXTEST	*WPSE*	*WPSE* / 0	*WPSE*	*~WPSE* / 1
WP_INTEST	*WPSE*	*~WPSE* / 1	*WPSE*	*WPSE* / 0
WP_INTEST_MBIST	-	0	-	0
WS_BYPASS	-	0	-	0
WP_BYPASS	0	1	0	1
WS_SAFE_SINGLE	0	1	0	1
WP_EXTEST_SEQ	*WPSE*	*WPSE* / 0	1	*~WPSE* / 1
WP_INTEST_SEQ	1	*~WPSE* / 1	*WPSE*	*WPSE* / 0

The *Scan enable* terminal serves as wrapper shift enable terminal during standard EXTEST and INTEST instructions. The WP_EXTEST_SEQ and WP_INTEST_SEQ instructions differ from the standard instructions only in their use of the scan enable signal. During external test the output cells that apply values are continuously kept in shift mode by a fixed logic '1' value. The same scenario is true for the input cells during internal test.

The *Hold enable* terminal is driven to a fixed logic '0' value by default. This enables functional operation. During the WP_BYPASS and WS_SAFE_SINGLE instructions, the hold path of the WBR cells is enabled to limit the amount of transitions in the test logic and the amount of current drawn by the wrapper.

During EXTEST instructions the WBR input cells need to capture data. This is achieved during serial instructions by assigning the inverted value of *CaptureWR* onto the *Hold enable* signal. During parallel EXTEST instructions, the capture operation is executed when *WPSE* is inactive. This means *Hold enable* values are identical to *WPSE* values. Since we decided to limit ourselves to standard scan test protocols for parallel instructions, the hold capability of the WBR cell is not used during these instructions. As a result, the *Hold enable* signal for input cells can alternatively be driven to a fixed logic '0' value as shown in Table 12. During EXTEST instructions, WBR output cells and WBR input cells will receive different control values because input cells are expected to behave differently from output cells during these instructions. One such difference in behavior is that, unlike WBR input cells, WBR output cells will not need to respond to the WBR capture event during EXTEST instructions.

During INTEST instructions, WBR input cells receive the same control as WBR output cells receive during EXTEST. Similarly, during INTEST, WBR output cells receive the same control as the WBR input cells receive during EXTEST. This shows that the inward and Outward Facing modes have mirrored functionality.

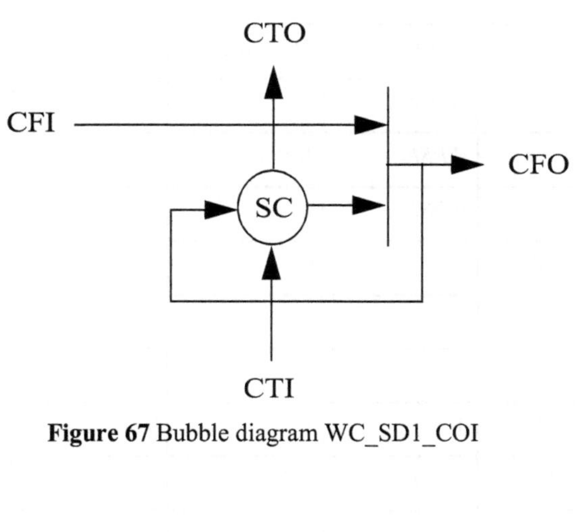

Figure 67 Bubble diagram WC_SD1_COI

6.6.3 Shared WBR cell design

All functional core terminals except CLK, RESET, READY and BC will be provided with a shared WBR cell. The WC_SF1_CII introduced in Section *6.3.3* (see page 143) will be used on these terminals. This cell is depicted in Figure 68 and the bubble diagram corresponding to this cell is shown in Figure 70 (page 161). This cell adds only one multiplexer in the functional path just like the multiplexing done in a traditional mux-scan cell implementation.

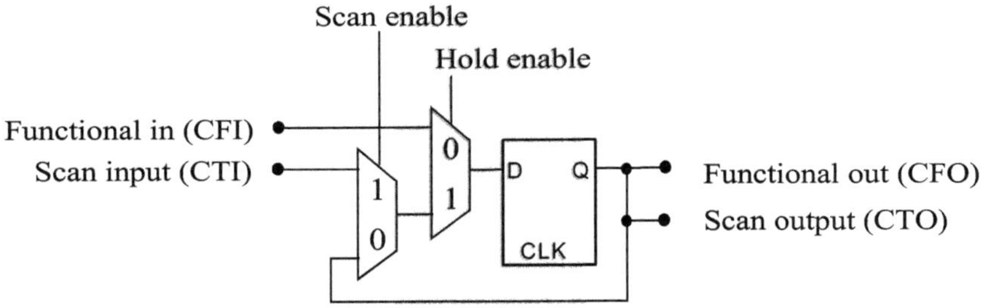

Figure 68 Implementation WC_SF1_CII

Table 13 Control values for WC_SF1_CII

	Input cell		*Output cell*	
Instruction	*Scan enable*	*Hold enable*	*Scan enable*	*Hold enable*
WS_EXTEST	*ShiftWR*	*~CaptureWR*	*ShiftWR*	1
WS_INTEST	*ShiftWR*	1	*ShiftWR*	*~CaptureWR*
WP_EXTEST	*WPSE*	*WPSE*	*WPSE*	1
WP_INTEST	*WPSE*	1	*WPSE*	*WPSE*
WP_INTEST_MBIST	-	0	-	0
WS_BYPASS	-	0	-	0
WP_BYPASS	0	1	0	1
WS_SAFE_SINGLE	0	1	0	1
WP_EXTEST_SEQ	*WPSE*	*WPSE*	1	1
WP_INTEST_SEQ	1	1	*WPSE*	*WPSE*

Table 13 shows values of signals that will control the shared WBR cell shown in Figure 68 during each EX wrapper instruction. Note that the hold enable columns for input and output cells in Table 12 and Table 13 are not exactly the same. For instance, in Table 12 WP_INTEST hold enable for the output cell can be WPSE or "0" and in Table 13 it is WPSE. So WPSE will control the hold enable signal in both cases. Dynamic control signals like scan enable often need to be treated like clocks during design. For this reason, static control may be preferred. In case static control is required an alternative shared WBR cell implementation can be used. This cell is presented in Figure 69 and uses exactly the same control signals as the dedicated WBR cell used for the READY and MBIST terminals. Note that this cell adds two muxes in the functional path and has not been selected for the EX core example for this reason.

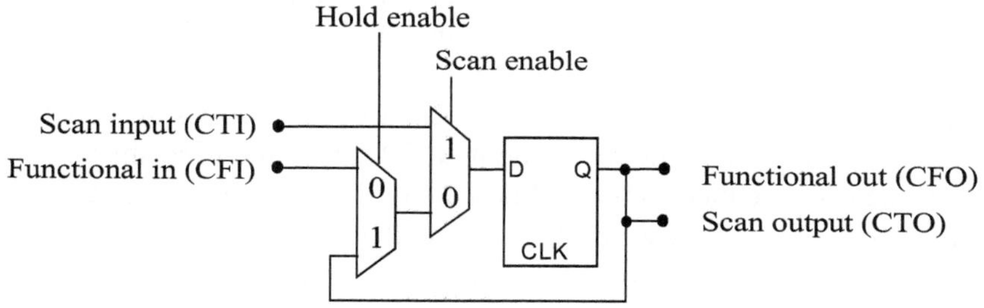

Figure 69 Example implementation WC_SF1_CII

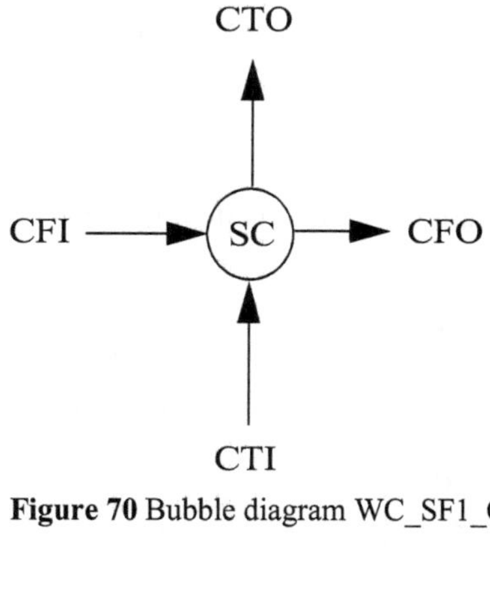

Figure 70 Bubble diagram WC_SF1_CII

6.6.4 Special WBR cell design

This section addresses isolation of terminals that are not mandated to have a WBR cell and/or require special behavior during test.

6.6.4.1 WBR cell for Bus Control

The BC terminal of the EX core needs special treatment during scan test as this terminal is required to output a stable value during the scan shift operation. The bus control function requires this in order for the bus external to the EX core to be in a known direction, during shift, to prevent contention. While the bus control signal is high, the bus allows drive from the external pins to the DIN pins of the core. While the bus control signal is low, the bus allows drive from the DOUT pins to the bus of the external logic.

The behavior required on the BC terminal can be obtained with a WBR cell that implements a silent shift path. A silent shift path is a shift path that does not disturb values on the CFO terminal of the WBR cell that it traverses. An example of such a cell is the WC_SD1_CII_UD cell shown in Figure 60 (page 147). This cell implements a silent shift path because it has an update stage preventing shift values from reaching its CFO terminal during shift.

For the EX core there is the additional requirement that scan testing via the parallel port can make use of standard ATPG and a standard test protocol. This means that no separate capture and update events can be supported. For this reason a customized WBR cell for the BC terminal will be implemented. This cell, shown in Figure 71, fits the characteristics of a WC_SD1_COI_G cell. The functional output of the cell can be driven by three sources. In functional operation data feeds from CFI to CFO unhindered. During scan shift operation the CFO output is fixed to logic value '1'. In addition, a *Bus disable* pin has been added to the WBR cell to statically drive the functional output to the safe logic '1' value during test.

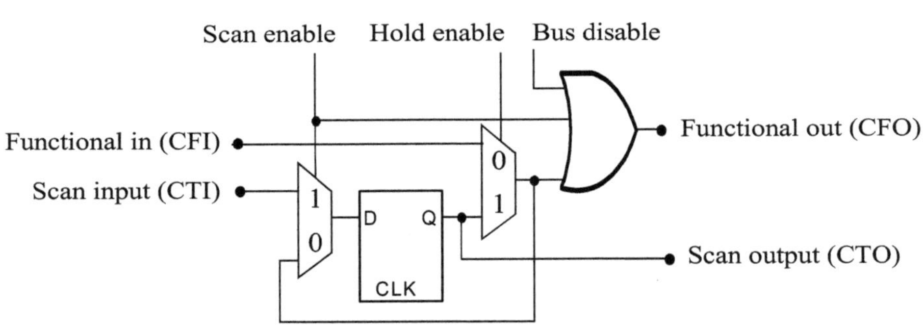

Figure 71 Implementation WC_SD1_COI_G for Bus Control

Table 14 shows control values for the WC_SD1_COI_G cell. The corresponding bubble diagram is depicted in Figure 72. Because the WC_SD1_COI_G cell is implemented by adding logic to the output of a WC_SD1_COI cell, both cells share the same control values. The additional *Bus disable* signal that forces a safe value from the WC_SD1_COI_G cell is activated while the wrapper is in Inward Facing mode to prevent disturbance to logic external to the wrapper. This safe behavior will also be activated by the WP_BYPASS and WS_SAFE_SINGLE instructions. During EXTEST instructions the scan enable signal will guarantee a safe output.

Table 14 Control values for WC_SD1_COI_G

	Output cell		
Instruction	*Scan enable*	*Hold enable*	*Bus disable*
WS_EXTEST	*ShiftWR*	1	0
WS_INTEST	*ShiftWR*	~*CaptureWR*	1
WP_EXTEST	*WPSE*	~*WPSE* / 1	0
WP_INTEST	*WPSE*	*WPSE* / 0	1
WP_INTEST_MBIST	-	0	1
WS_BYPASS	-	0	0
WP_BYPASS	0	1	1
WS_SAFE_SINGLE	0	1	1
WP_EXTEST_SEQ	1	~*WPSE* / 1	0
WP_INTEST_SEQ	*WPSE*	*WPSE* / 0	1

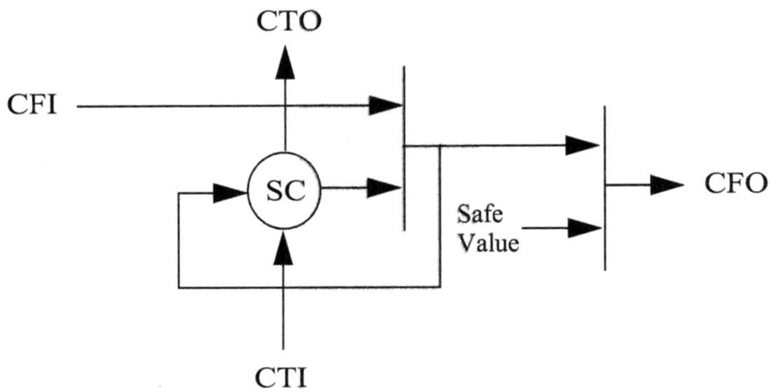

Figure 72 Bubble diagram WC_SD1_COI_G

6.6.4.2 WBR cell for Reset

For the EX core it was decided that a WBR cell will be attached to the RESET pin, though it is not required by the 1500 standard. Direct control of the RESET is standard practice during internal test mode. However, there may be logic external to the core that needs observation to get high test coverage during scan test of external logic. An example of this type of logic is shown in Figure 73.

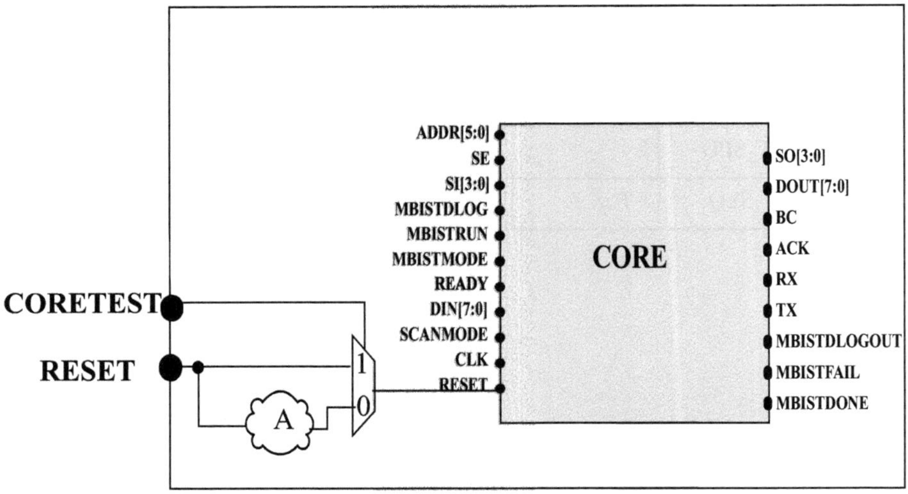

Figure 73 External Reset Logic Example

The WBR cell that was designed for the Bus Control output of the EX core can also be used to isolate the asynchronous RESET input of the EX core.

This cell can control the RESET input to prevent asynchronous resets during synchronous test operation, in a way similar to controlling the bus control output to prevent contention.

One of the requirements for the EX core is that standard (non-sequential) ATPG must be used for test pattern generation. This means, for this example, that the asynchronous RESET signal must be directly controlled from a top level pin to meet the ATPG requirements. During external test mode, the logic labeled A in Figure 73 can be tested provided there is a WBR cell to observe the output of the logic. Therefore, an observe only WBR cell is provided for the RESET signal on the EX core. A WC_SD1_CII_O cell shown in Figure 74 will be used. The corresponding bubble diagram is depicted in Figure 75 (page 167). The cell's functional input signal is directly driven through to the core logic, and can also be observed in the cell's storage element. This WBR cell meets the requirements identified above for the EX core RESET terminal.

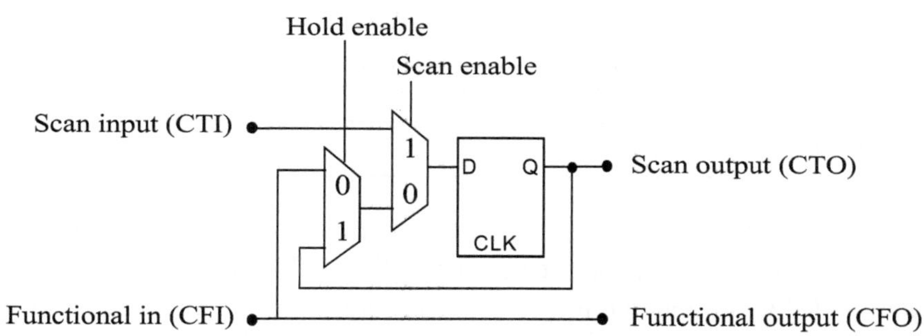

Figure 74 Implementation WC_SD1_CII_O

Table 15 Control values for WC_SD1_CII_O

	Input cell	
Instruction	*Scan enable*	*Hold enable*
WS_EXTEST	*ShiftWR*	*~CaptureWR*
WS_INTEST	*ShiftWR*	-
WP_EXTEST	*WPSE*	*WPSE* / 0
WP_INTEST	*WPSE*	-
WP_INTEST_MBIST	-	-
WS_BYPASS	-	-
WP_BYPASS	0	1
WS_SAFE_SINGLE	0	1
WP_EXTEST_SEQ	*WPSE*	*WPSE* / 0
WP_INTEST_SEQ	1	-

The WC_SD1_CII_O cell has reduced functionality compared to a full-featured cell. The functional path from CFI to CFO does not pass any logic gate and thus is unconditioned. The cell has no drive capability, so is not involved in any INTEST instruction apart from shifting. Furthermore, Table 15 shows that the observe-only cell contributes to test during EXTEST execution only. In line with all other WBR cells, the WC_SD1_CII_O storage element holds its contents during the WP_BYPASS and WS_SAFE_SINGLE instructions to avoid unnecessary transitions.

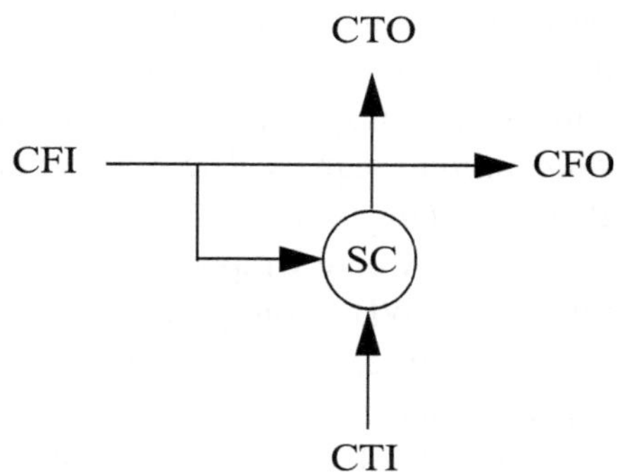

Figure 75 Bubble diagram WC_SD1_CII_O

6.6.5 WBR design

Now that all of the WBR cells that will be employed have been described, the WBR can be put together. Table 16 compiles the set of control signals required for the WBR. This information will be critical to the design of the WIR circuitry since these control signals are expected to come from the WIR as shown in Section *8.3.3* (see page 203).

Table 16 Control values for the WBR chain

	Input cells	*Output cells*	*Input cells*	*Output cells*	*BC cell*
Instruction	wse_inputs	wse_outputs	hold_inputs	hold_outputs	bus_ disable
WS_EXTEST	*ShiftWR*	*ShiftWR*	*~CaptureWR*	1	0
WS_INTEST	*ShiftWR*	*ShiftWR*	1	*~CaptureWR*	1
WP_EXTEST	*WPSE*	*WPSE*	*WPSE*	1	0
WP_INTEST	*WPSE*	*WPSE*	1	*WPSE*	1
WP_INTEST_MBIST	0	0	0	0	1
WS_BYPASS	0	0	0	0	0
WP_BYPASS	0	0	1	1	1
WS_SAFE_SINGLE	0	0	1	1	1
WP_EXTEST_SEQ	*WPSE*	1	*WPSE*	1	0
WP_INTEST_SEQ	1	*WPSE*	1	*WPSE*	1

In addition to WRCK and CLK, two signals are needed to control all WBR input cells; the *wse_inputs* signal which connects to the wrapper cell Scan enable terminals and the *hold_inputs* signal which connects to the Hold enable of the same cells. In a similar way, the WBR output cells are connected to the *wse_outputs* and *hold_outputs* signals. The wrapper cell for the BC terminal also receives the dedicated *bus_disable* signal.

A schematic overview of the EX core wrapper is given in Figure 76 (page 169). This figure shows the single WBR configuration that is used during the mandated serial WS_EXTEST instruction. The WBR cells are represented by their bubble diagrams and connected in a single shift path between the wbr_si and wbr_so signals. These signals feed into WSI and WSO as shown in Figure 104 (page 214).

It can be derived from the bubble diagrams that all, except eight, WBR cells reuse a functional storage element. The wrapper terminals that do not have a WBR cell are the functional clock CLK and WPP terminals that consist of scan-in and scan-out terminals WPSI, WPSO, and scan enable terminal WPSE.

6.6.5.1 Parallel configuration of the WBR

The WBR is divided into multiple segments during parallel EXTEST and INTEST instructions. This parallel configuration of the WBR is shown in Figure 77 (page 170). The number of EX core WBR segments equals the width of the EX core parallel data port (WPP) which is four. The WBR cells are equally distributed over the segments to limit the amount of scan cycles needed to access all cells. The scan length for the parallel configuration of the WBR is defined by the longest segment, which is nine for the EX core (lengths of segments 2 and 3). The WBR segments are used in multiple register configurations that depend on the active instruction. This is further explained in Section *8.3.4* (see page 211).

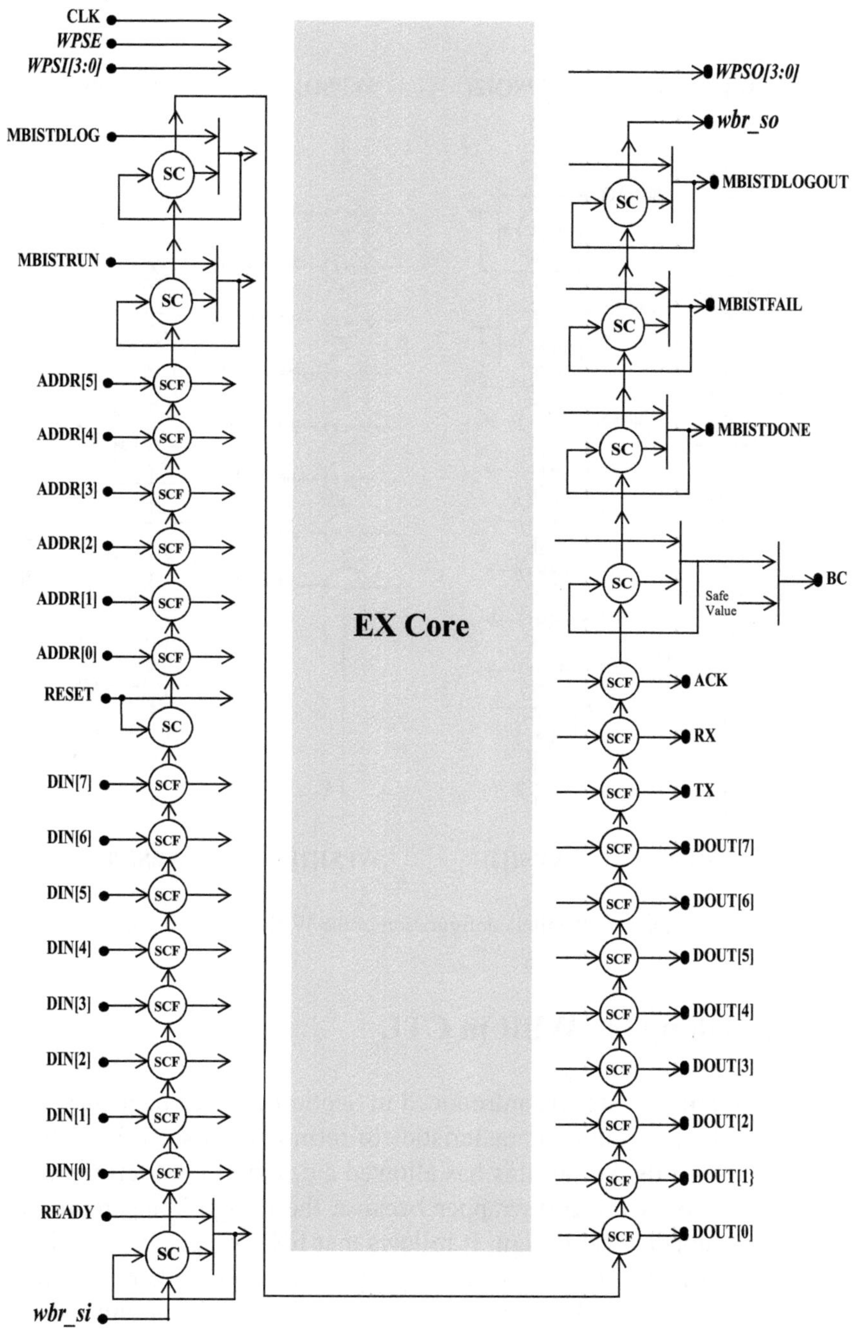

Figure 76 WBR for the EX core

Figure 77 shows parallel configuration of the WBR.

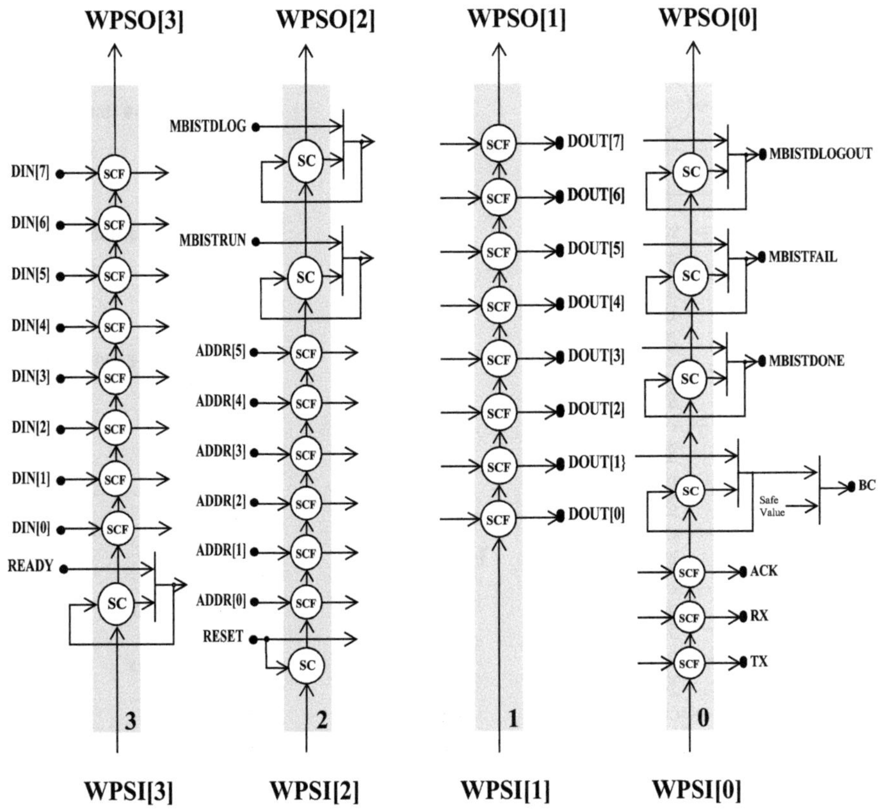

Figure 77 Parallel configuration of the WBR for the EX core

6.7 Describing the WBR in CTL

The **Internal** block was introduced in Section *4.2* (see page 38) as a block of CTL code describing characteristics of terminals as seen from the design boundary, inwards. So far, this has allowed a description of terminals such as WSP terminals of the EX wrapper because the design being represented in CTL is a wrapper-level design. It follows that EX wrapper WBR cells should also be described in the **Internal** block because these cells reside inside the wrapper. However, CTL code may be done at the core level rather than at the wrapper level in cases of unwrapped compliant cores. In such cases, the design level being described in CTL is the core level and use of the **Internal** block to describe wrapper components such as WBR cells would be inaccurate since these cells would exist outside of the design level being represented in CTL (i.e. the core). In these cases use of the CTL **External** block is required as this block is meant for describing logic residing outside the design level being described in CTL. Both internal and external description scenarios will be used in the following paragraphs to illustrate description of WBR cells in CTL.

6.7.1 Describing WBR cells in CTL at a Wrapper Level

As presented above, wrapper level description of WBR cells is done inside the **Internal** block which resides inside the **CTLMode** construct as seen in Section *4.2* (see page 38). The following CTL syntax was defined by 1149.6 to accommodate this description.

```
<6.1>   Environment {
<6.2>       CTLMode {
<6.3>           Internal {
<6.4>               SIG_NAME {
<6.5>                   (IsConnected <In|Out> (<Direct | Indirect>) {
<6.6>                       StateElement <Scan | NonScan> (cellref_expr) ; )
<6.7>                   Wrapper <IEEE1500 | None | User USER_DEFINED >
<6.8>                       (< CellID cell_enum);
<6.9>                   } // IsConnected block
<6.10>              } // SIG_NAME block
<6.11>          } // Internal block
<6.12>      } // CTLMode block
<6.13>  } // Environment block
```

where:

- SIG_NAME is the wrapper level signal for which connectivity information is being provided.
- **IsConnected** begins the description of connectivity information where one end of the connection (source or sink of the connection) is a design interface terminal and the other end (source or sink of the connection) is inside the design being described.
- **In** describes that SIG_NAME is the source of the connection being described.
- **Out** describes that SIG_NAME is the sink of the connection being described.
- **Direct** signifies a sensitized direct connection equivalent to a wire. This is the default type of connection when neither **Direct** nor **Indirect** is specified.

- **Indirect** signifies a non-sensitized connection that requires sensitization.
- **StateElement** describes a storage element that is connected to SIG_NAME.

  ▪*cellref_expr* names the **Scan** or **NonScan** storage element that is represented by the **StateElement** keyword. This name must match one of the WBR cell names defined in the **ScanStructure** block representing the WBR chain as defined in Section *6.8* (see page 174).
- **Scan** represents connection to a state element that is part of a scan chain visible in the current CTLMode block. One way to make scan chains visible in the current CTLMode block is through the inheritance feature introduced in Section *4.2* **(see page 38)**.
- **NonScan** represents connection to a storage element that is not defined as part of any scan chains visible in the current CTLMode block.
- **Wrapper IEEE 1500** signifies that the connection is done through a 1500 wrapper cell. The **None** argument of the wrapper keyword is the default and signifies that no wrapper cell is provided. The **User** argument is reserved for non-1500 compliant wrapper cells and therefore will not be described is this book.
- **CellID** signifies that the identity of a wrapper cell is being provided.

  ▪*cell_enum* represents the 1500 wrapper cell type used in the connection being described. 1500 compliant cell types are described in Section *6.4.1.2* (see page 149).

6.7.2 Describing WBR Cells in CTL at a Core Level

1450.6 syntax for core level description of WBR cells follows.

```
<6.14>  Environment{
<6.15>      CTLMode {
<6.16>          External {
<6.17>              SIG_NAME {
<6.18>                  ConnectTo {
<6.19>                  Wrapper <IEEE1500 | None | User USER_DEFINED >
<6.20>                  (< CellID cell_enum | PinID USER_DEFINED_PIN_ID >);
<6.21>                  } // ConnectTo block
<6.22>              } // SIG_NAME block
<6.23>          } // External block
<6.24>      } // CTLMode block
<6.25>  } // Environment block
```

Where:

- **ConnectTo** defines connectivity to an endpoint that exists outside the design level being described in CTL. If the CTL describes an unwrapped compliant core, and the core level terminal being described here is meant to be connected to a WBR cell, then the **ConnectTo** statement must be used to specify connection to an example WBR cell per 1500 requirements as described in Section *1.2.1* (see page 7).
- **PinID** keyword begins the description of a user-defined pin attribute. The **PinID** keyword will not be used in this book.
  - USER_DEFINED_PIN_ID represents a user-defined attribute meant to characterize one or more pins.

With the exception **PinID**, all arguments of the **Wrapper** keyword have already been defined in Section *6.7.1* (see page 171).

6.7.3 Describing Scan Chains in CTL

CTL scan chains are defined using the **ScanStructures** construct. This construct originated from STIL and is part of STIL information natively inherited by CTL. STIL defines syntax of the **ScanStructures** construct as follows:

```
<6.26>  ScanStructures (scanstructure_block_name) {
<6.27>      ScanChain scanchain_bloc_name {
<6.28>          ScanLength integer;
<6.29>          ScanCells list_of_scancells_names;
<6.30>          ScanIn signal;
<6.31>          ScanOut signal;
<6.32>          ScanMasterClock signal_list;
<6.33>          ScanSlaveClock signal_list;
<6.34>          ScanInversion 0_or_1
<6.35>      } // ScanChain block
<6.36>  } // ScanStructures block
```

where argument of the **ScanInversion** keyword represents inversion or non-inversion of the scan chain with respect to the scan input and scan output pins. Other keywords listed above seem to not require further clarification. The **ScanChain** block accepts additional keywords that are not shown here but can be obtained from the 1450.0 document.

6.8 Describing the WBR chain of the EX wrapper in CTL

From a CTL description perspective, the WBR is a scan chain to be described with the **ScanStructures** keyword defined in Section *6.7.3* (see page 173). In particular, 1500 requires use of the **ScanLength** and **ScanCells** keywords of the **ScanStructures** block in the definition of the WBR chain.

Note: Rule 17.1.1(c) of the 2005 version of the 1500 standard wrongly refers to the use of a Length keyword to express scan chain length. The correct keyword to use is **ScanLength** as stated here.

6.8.1 EX wrapper serial mode WBR chain

The **ScanStructures** block corresponding to the EX wrapper follows:

```
<6.37>  ScanStructures EX_wrapper_WSP_chains {
<6.38>      ScanChain EX_WBR_WSP_chain {
<6.39>          ScanLength 33;
<6.40>          ScanIn WSI;
<6.41>          ScanOut WSO;
<6.42>          ScanMasterClock WRCK;
<6.43>          ScanCells {
<6.44>  WBR_READY; WBR_DIN[0..7]; WBR_RESET; WBR_ADDR[0..5];
          WBR_MBISTRUN; WBR_MBISTDLOG; WBR_DOUT[0..7];WBR_TX;
          WBR_RX; WBR_ACK; WBR_BC; WBR_MBISTDONE;
          WBR_MBISTFAIL; WBR_MBISTDLOGOUT;
<6.45>          } // ScanCells block
<6.46>      } // ScanChain block
<6.47>  } // ScanStructures
```

Where:

- EX_wrapper_WSP_chains is the name of a **ScanStructures** block containing EX wrapper scan chain definitions, and EX_WBR_WSP_chain is the name of the WBR scan chain in serial mode.
- WBR cell instance names were formed by adding the prefix "WBR_" to the name of the corresponding EX core terminals. The resulting mapping of the core terminal name to WBR instance name is shown in Table 17 This mapping results from EX core WBR cell selection discussed in section Section *6.6.2.1* (see page 157). The scan cell instance names follow the scan-input to scan-output order represented in Figure 76 (page 169).

6.8.2 EX wrapper parallel mode WBR chains

EX wrapper WBR chains that are active in parallel mode will be defined with a dedicated **ScanStructures** block as follows:

```
<6.48>  ScanStructures EX_wrapper_WPP_chains {
<6.49>      ScanChain EX_WBR_WPP_chain_0 {
<6.50>          ScanLength 7;
<6.51>          ScanIn WPSI[0];
<6.52>          ScanOut WPSO[0];
<6.53>          ScanMasterClock WRCK; CLK;
<6.54>          ScanCells {
<6.55>  WBR_TX; WBR_RX; WBR_ACK; WBR_BC; WBR_MBISTDONE;
           WBR_MBISTFAIL; WBR_MBISTDLOGOUT;
<6.56>          } // ScanCells block
<6.57>      } // ScanChain block
<6.58>      ScanChain EX_WBR_WPP_chain_1 {
<6.59>          ScanLength 8;
<6.60>          ScanIn WPSI[1];
<6.61>          ScanOut WPSO[1];
<6.62>          ScanMasterClock WRCK; CLK;
<6.63>          ScanCells {
<6.64>              WBR_DOUT[0..7];
<6.65>          } // ScanCells block
<6.66>      } // ScanChain block
<6.67>      ScanChain EX_WBR_WPP_chain_2 {
<6.68>          ScanLength 9;
<6.69>          ScanIn WPSI[2];
<6.70>          ScanOut WPSO[2];
<6.71>          ScanMasterClock WRCK; CLK;
<6.72>          ScanCells {
<6.73>              WBR_RESET; WBR_ADDR[0..5]; WBR_MBISTRUN;
                   WBR_MBISTDLOG;
<6.74>          } // ScanCells block
<6.75>      } // ScanChain block
<6.76>      ScanChain EX_WBR_WPP_chain_3 {
<6.77>          ScanLength 9;
<6.78>          ScanIn WPSI[3];
<6.79>          ScanOut WPSO[3];
<6.80>          ScanMasterClock WRCK; CLK;
<6.81>          ScanCells {
<6.82>              WBR_READY; WBR_DIN[0..7];
<6.83>          } // ScanCells block
<6.84>      } // ScanChain block
<6.85>  } // ScanStructures
```

The above parallel mode scan chains correspond to the parallel mode WBR implementation shown in Figure 77 (page 170).

6.9 Describing WBR Cells of the EX Wrapper in CTL

Table 17 shows WBR cell type mapping to EX core terminal names. In this table, EX core terminals without provision of a WBR cell have "N/A" in their corresponding WBR cell name field.

Table 17 WBR cell name and type mapping to EX wrapper terminals

EX core terminal name	*WBR cell instance name*	*WBR cell type*
READY	WBR_READY	WC_SD1_COI
MBISTDLOG	WBR_MBISTDLOG	WC_SD1_COI
MBISTRUN	WBR_MBISTRUN	WC_SD1_COI
MBISTDLOGOUT	WBR_MBISTDLOGOUT	WC_SD1_COI
MBISTDONE	WBR_MBISTDONE	WC_SD1_COI
MBISTFAIL	WBR_MBISTFAIL	WC_SD1_COI
ADDR[5:0]	WBR_ADDR[5:0]	WC_SF1_CII
DIN[7:0]	WBR_DIN[7:0]	WC_SF1_CII
DOUT[7:0]	WBR_DOUT[7:0]	WC_SF1_CII
ACK	WBR_ACK	WC_SF1_CII
RX	WBR_RX	WC_SF1_CII
TX	WBR_TX	WC_SF1_CII
RESET	WBR_RESET	WC_SD1_CII_O
BC	WBR_BC	WC_SD1_COI_G
CLK	N/A	N/A
MBISTMODE	N/A	N/A
SCANMODE	N/A	N/A
SE	N/A	N/A
SI[3:0]	N/A	N/A
SO[3:0]	N/A	N/A

The CTL code describing terminals provisioned with WBR cells is shown below.

```
Environment EX_wrapper {
    CTLMode {
        Internal {
            READY {
                IsConnected In {
                    StateElement Scan WBR_READY;
                    Wrapper IEEE1500 CellID WC_SD1_COI;
                }
            } // READY block
            MBISTDLOG {
                IsConnected In {
                    StateElement Scan WBR_MBISTDLOG;
                    Wrapper IEEE1500 CellID WC_SD1_COI;
                }
            } // MBISTDLOG block
            MBISTRUN{
                IsConnected In {
                    StateElement Scan WBR_MBISTRUN;
                    Wrapper IEEE1500 CellID WC_SD1_COI;
                }
            } // MBISTRUN block
            MBISTDLOGOUT {
                IsConnected In {
                    StateElement Scan WBR_MBISTDLOGOUT;
                    Wrapper IEEE1500 CellID WC_SD1_COI;
                }
            } // MBISTDLOGOUT block
            MBISTDONE {
                IsConnected In {
                    StateElement Scan WBR_MBISTDONE;
                    Wrapper IEEE1500 CellID WC_SD1_COI;
                }
            } // MBISTDONE block

            MBISTFAIL {
                IsConnected In {
                    StateElement Scan WBR_MBISTFAIL ;
                    Wrapper IEEE1500 CellID WC_SD1_COI;
                }
            } // MBISTFAIL block
```

```
ADDR[0..5] {
    IsConnected In {
        StateElement Scan WBR_ADDR[0..5];
        Wrapper IEEE1500 CellID WC_SF1_CII;
    }
} // ADDR[0..5] block
DIN[0..7]{
    IsConnected In {
        StateElement Scan WBR_DIN[0..7];
        Wrapper IEEE1500 CellID WC_SF1_CII;
    }
} // DIN[0..7] block
DOUT[0..7] {
    IsConnected In {
        StateElement Scan WBR_DOUT[0..7];
        Wrapper IEEE1500 CellID WC_SF1_CII;
    }
} // DOUT[0..7] block
TX {
    IsConnected In {
        StateElement Scan WBR_TX;
        Wrapper IEEE1500 CellID WC_SF1_CII;
    }
} // TX block
ACK {
    IsConnected In {
        StateElement Scan WBR_ACK;
        Wrapper IEEE1500 CellID WC_SF1_CII;
    }
} // ACK block
RX {
    IsConnected In {
        StateElement Scan WBR_RX;
        Wrapper IEEE1500 CellID WC_SF1_CII;
    }
} // RX block
RESET {
    IsConnected In {
        StateElement Scan WBR_RESET;
        Wrapper IEEE1500 CellID WC_SD1_CII_O;
    }
} // RESET block
```

```
<6.168>                 BC {
<6.169>                     IsConnected In {
<6.170>                         StateElement Scan WBR_BC;
<6.171>                         Wrapper IEEE1500 CellID WC_SD1_COI_G;
<6.172>                     }
<6.173>                 } // BC block
<6.174>             } // Internal block
<6.175>         } // CTLMode block
<6.176>     } // Environment block
```

Note that EX core CTL description is contained inside an **Internal** block as opposed to being contained inside an **External** block. This was explained in section Section *6.7* (see page 170).

Chapter 7
Design of the WBY

7.1 Introduction

The Wrapper Bypass Register (WBY) is utilized to bypass other Wrapper Data Registers (WDRs) and is a required element of the 1500 wrapper. The WBY is a mandatory dedicated test register and can be accessed only through the WSP signals. This register must be clocked by WRCK.

In an SOC, the WBRs of multiple cores can be concatenated. This may be useful in order to implement a minimal number of test ports. Rather than one or more scan inputs and outputs going to and coming from each WBR to SOC pins, a single scan in (WSI) and scan out (WSO) can be used if all of the WBRs are concatenated. However, this could create a very long scan chain. Any WBR that is not required during a specific test can be bypassed with the WBY so that the concatenated wrapper chain becomes shorter, thus reducing test time.

7.2 WBY Construction

The WBY consists of one or more registers like the one shown in Figure 78 (page 182). 1500 describes only the behavior of this register, not its construction. Even though the symbol for a flip-flop is shown in Figure 78, the WBY behavior could also be constructed from latches. The standard also mandates that the WBY must retain its shift values in the absence of active WBY operations. To meet this requirement, clock gating logic can be added to the WBY clock input to turn off WRCK.

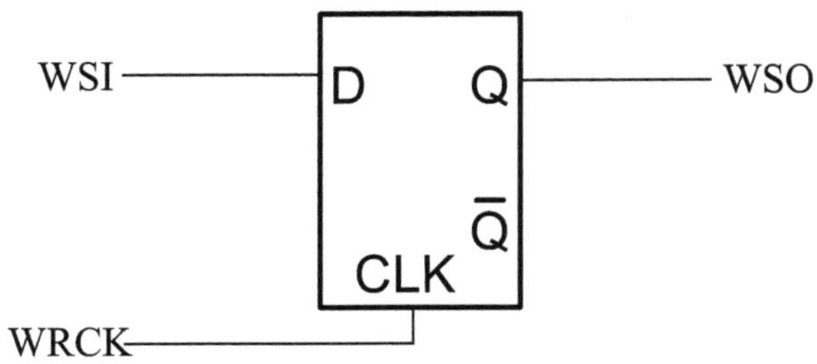

Figure 78 WBY Example

The 1500 wrapper contains logic that selects either a WDR or the WBY path as shown in Figure 79.

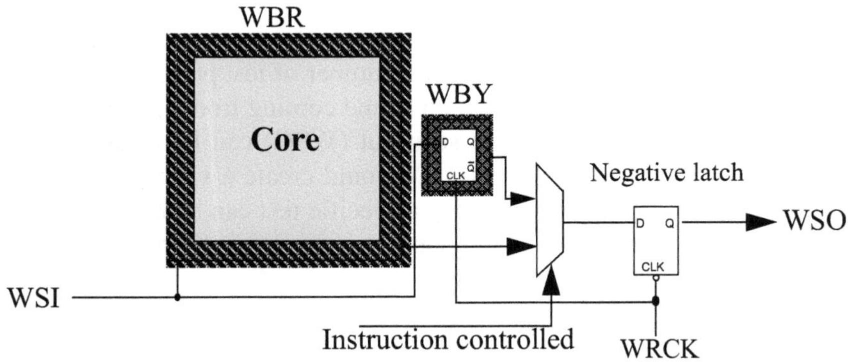

Figure 79 WBY mechanism

WSI is connected to the D port of the example WBY and WRCK is connected to the CLK port. The output must feed into WSO. These three WSP signal connections are shown in Figure 78. As can be seen in Figure 79, the WSI goes directly to the WBR and the WBY. These signals can be gated and controlled by the WIR to reduce power and noise during test as shown in Figure 80.

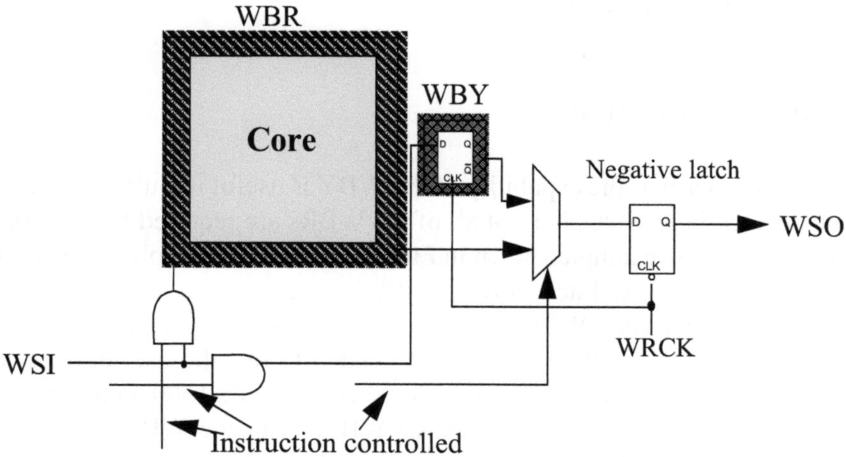

Figure 80 WBY mechanism with gated WSI

Many different instructions can configure the wrapper to utilize the WBY path. Per 1500 mandate, there must be a register between WSI and WSO during all instructions. If a WDR is not connected between WSI and WSO during an instruction, the WBY is used to fulfill this requirement. It is intuitive that the WBY path is employed during the WS_BYPASS instruction as described in Section *5.1.1.1* (see page 80). However, other instructions that do not use a WDR register or specifically the WBY register must also configure the WBY register in the WSI to WSO path. An example of this is the WS_SAFE instruction as shown in Figure 81. This instruction simply gates the outputs of WBR cells, and therefore does not need to explicitly shift any data through the WBR. In this scenario, the WBY is used as the conduit between WSI and WSO.

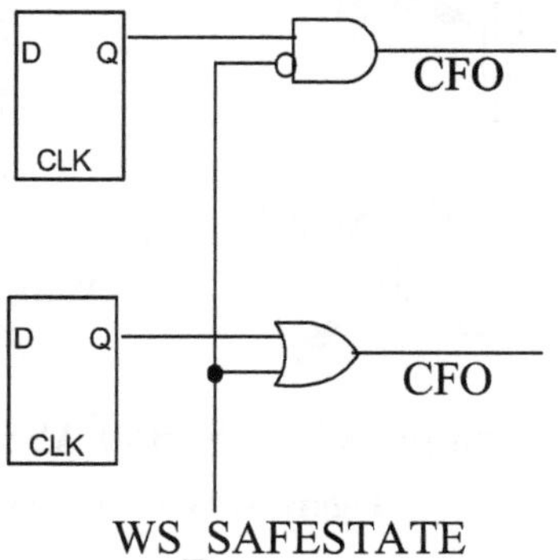

Figure 81 WS_SAFE example

WS_CLAMP is another 1500 standard instruction that implements the WBY between WSI and WSO.

7.3 WBY Utilization

As stated earlier, the capability of the WBY is useful if multiple WBRs are concatenated in an SOC, but not all of the WBRs are required for a particular test. Consider the example shown in Figure 82. In this example, there are four 1500 wrapped cores. Each core's WBR is concatenated to another core's WBR so that all four WBRs become a single scan chain. However, only Core1 and Core2's 1500 wrappers are needed to test the user-defined logic shown in Figure 82. In this scenario, Core3 and Core4's 1500 wrappers can be programmed to bypass their respective WBRs with the WBY (e.g. by programming the WS_BYPASS instruction into the WIR).

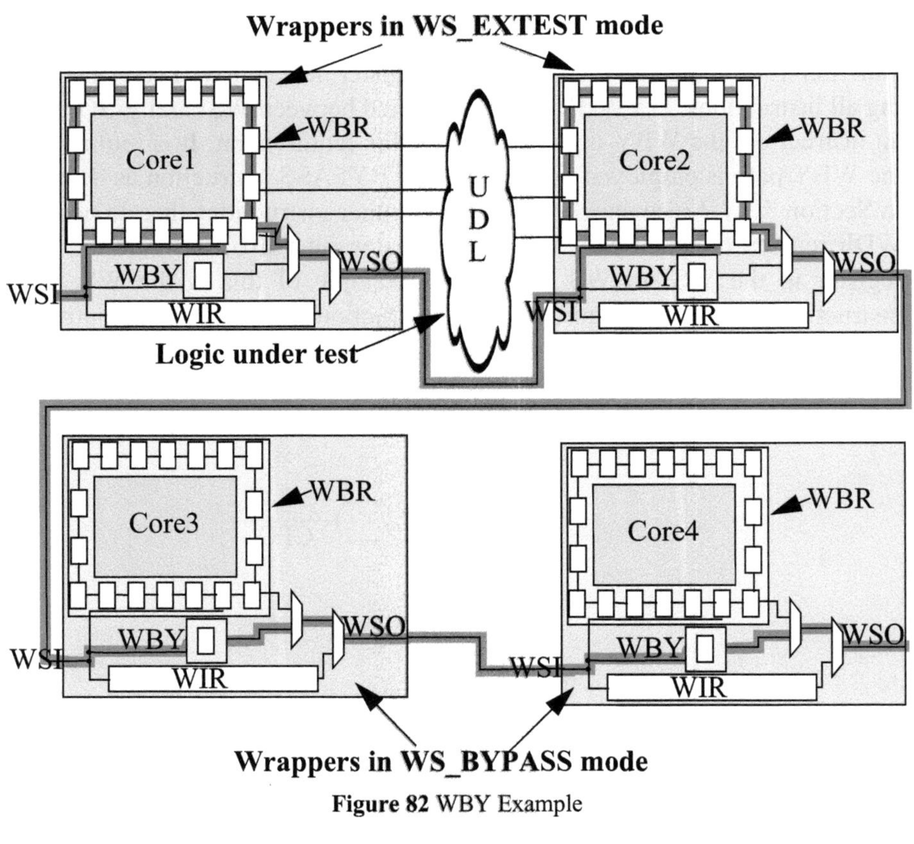

Figure 82 WBY Example

7.4 WBY for the EX Core

The WBY shown in Figure 83 is the WBY that will be used on the EX core to help create the fully wrapped EX core.

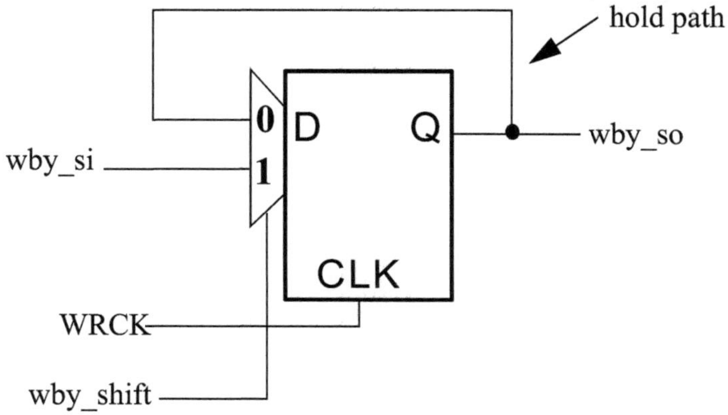

Figure 83 EX Core WBY register

The wby_shift signal that enables shift of the WBY register is derived from the WIR described in Section *8.3.3.2* (see page 203). In the absence of a shift operation (wby_shift set to 0) the EX core WBY register shown in Figure 83 will hold its shift value through the hold path. This is part of the WBY hold behavior required by the 1500 standard.

7.5 Describing the WBY in CTL for the EX Core

From a CTL perspective the EX core WBY is a scan chain of a single element and therefore will be defined using the **ScanStructures** syntax defined in Section *6.7.3* (see page 173). The resulting CTL description follows:

```
<7.1>   ScanStructures EX_wrapper_WSP_chains {
<7.2>           ScanChain EX_WBY_chain {
<7.3>               ScanLength 1;
<7.4>               ScanIn WSI;
<7.5>               ScanOut WSO;
<7.6>               ScanCells wby_cell;
<7.7>           } // ScanChain block
<7.8>   } // ScanStructures block
```

Where:

- EX_wrapper_WSP_chains is a **ScanStructure** group containing all EX wrapper scan chains
- EX_WBY_chain is the name of the EX wrapper WBY chain
- "wby_cell" is the WBY register instance name.

Chapter 8
Design of the WIR

The 1500 wrapper contains a standard mechanism for handling test control. This mechanism is called the Wrapper Instruction Register (WIR) and is primarily meant to configure the WBR. However, it also controls the mode of the core embedded within the 1500 wrapper. Wrapper instructions are serially loaded through the standard WSP into the WIR.

The following types of test control signals are distinguished during test:

- **static test control signals**: signals that set up the static conditions for a test, and remain stable during the entire test. Test modes or configurations are activated by static test control signals.
- **dynamic test control signals**: signals that change value even during the execution of one test pattern.

Static test control signals are controlled by the WIR.

8.1 WIR architecture

A shift register control mechanism has been chosen for the WIR because of the ability to daisy chain an arbitrary number of these control structures. This allows the flexibility required when producing a core-based design that has multiple wrappers embedded. The basic structure of the WIR is shown in Figure 84 (page 188). This consists of a shift register and an update register with instruction decoding logic. This structure resembles the registers used in 1149.1.

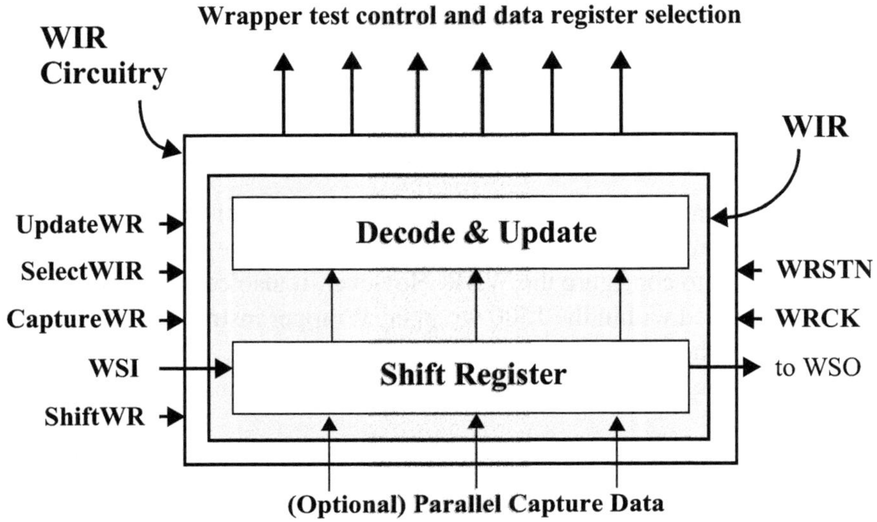

Figure 84 Example WIR

The WDR control signals and the core's active test mode are driven by the update register content. The update register is typically built of memory elements that are asynchronously reset when the active low WRSTN signal is asserted. As a result of this assertion, the wrapper goes into a disabled state which enables functional operation of the embedded core. The shift register allows the loading of a new instruction without disturbing the active test mode of the wrapper.

Note that the number of update registers could outnumber the number of state elements in the shift register if there is decode logic between the shift and update logic.

8.1.1 WIR Operation

Operation of the WIR is controlled by the WSC terminals of the wrapper. While loading instructions into the wrapper, a standard protocol that activates WIR operations, one at a time and in the correct order, needs to be followed.

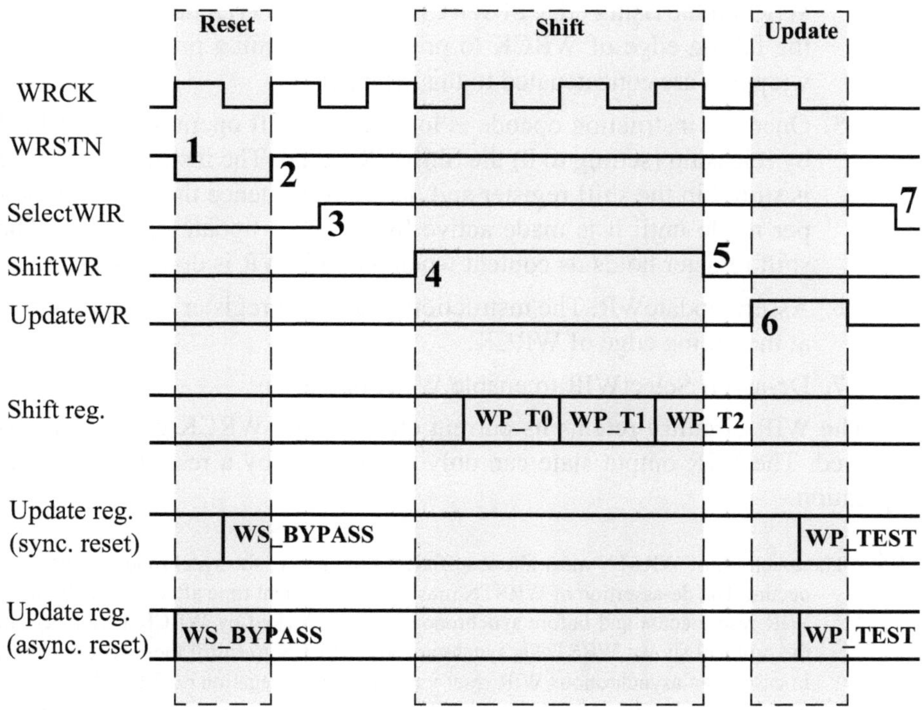

Figure 85 Protocol for resetting the WIR and loading an instruction

The recommended protocol is illustrated, including wrapper reset, in Figure 85. The protocol ensures that at least one complete WRCK clock pulse is applied when WRSTN is enabled. This ensures that both synchronous and asynchronous reset implementations will maintain the WS_BYPASS mode when the wrapper is enabled. For further details see Section *8.2.3* (see page 191).

The protocol shown in Figure 85 is as follows:

1. Reset is activated, the wrapper is disabled and the core is put into functional mode.

   a. If reset is synchronous, the update register of a wrapper is loaded with WS_BYPASS on the negative edge of WRCK.

2. Reset is de-activated, the wrapper is enabled.

3. If SelectWIR is not enabled, it must be enabled next.

4. Assert ShiftWR to enable the shift of instruction opcodes into the WIR shift register. Instruction opcode data is loaded from WSI to WSO on the rising edge of WRCK. The WSO serial output changes at the falling edge of WRCK to prevent hold timing problems if other wrappers are concatenated to this wrapper.
5. Once the instruction opcode is loaded, the shift operation is disabled by releasing (setting to 0) the ShiftWR signal. The instruction opcode is stored in the shift register and does not influence the current wrapper mode until it is made active by the WIR update operation. The shift register holds its content when the ShiftWR is de-asserted.
6. Assert UpdateWR. The instruction in the shift register becomes active at the falling edge of WRCK.
7. De-assert SelectWIR to enable WDR operation.

The WIR circuitry retains its current output if the WRCK clock signal is stopped. The WIR output state can only be changed by a reset or an update operation.

Note: The assertion of WRSTN must allow sufficient time before the synchronous WIR reset occurs. The de-assertion of WRSTN must allow sufficient time after the synchronous WIR reset occurs and before synchronous events clocked by WRCK take place. In the protocol shown WRSTN is synchronized to WRCK to fulfill these requirements. In case of an asynchronous WIR reset implementation negation of WRSTN need not be timing critical if sufficient time is allowed before synchronous events take place.

8.2 Understanding the WIR Rules

This section gives the reasoning behind the WIR rules defined in the 1500 standard document. It also covers the Wrapper Enabled and Disabled states since these states are closely coupled with the WIR and its implementation.

8.2.1 WIR Configuration and DR selection

Only one WIR shift chain, controlled by the WSC signals, is allowed per core. This does not mean that all WIR circuitry must reside in one design entity. Multiple WIR shift chains from different design entities can be concatenated to create a single chain within a core. This will look like a single WIR and thus be compliant to the 1500 standard. This type of concatenation could occur for a 1500 wrapped core that has another 1500 wrapped core embedded as discussed in Chapter 9.

Since the WIR defines the active mode of a design, it must be possible to access the WIR independently of this active mode to prevent deadlock situations. The WIR is expected to be accessible unconditionally by the WSP when SelectWIR is asserted. This provides a simple 'plug-and-play' wrapper interface. WSI is used for loading instructions into the WIR and for loading data into wrapper data registers during serial instructions.

While SelectWIR is deasserted, the register between the WSI and WSO terminals is determined by the active instruction in the WIR. All standard instructions predefine which register resides between WSI and WSO. However, user-defined instructions must also follow this rule. If no specific register is required between WSI and WSO for a given instruction, then the WBY must reside in this space. This guarantees serial access to other wrappers in an SOC, if those wrappers have been concatenated to the wrapper under discussion.

8.2.2 Wrapper Disabled and Enabled States

There are two wrapper states; enabled and disabled. During functional mode, the wrapper is disabled or in transparent mode. When WRSTN is enabled (forced to 0), it unconditionally forces the wrapper into its disabled state and the core into functional mode by implementing the WS_BYPASS instruction. Since the WIR and its circuitry fully define the configuration of the WDRs and the mode of the embedded core, its implementation can fulfill this behavior.

The most straightforward implementation of this behavior is done by asynchronously resetting the update register. However, a synchronous implementation is also allowed and will be discussed in the next section.

8.2.3 WIR Design

The WIR is composed of dedicated test logic. The main reasons for this are

- the need for modes in which functional operation can continue unhindered while the control mechanism is active.
- the fact that functional logic must be testable once a static test mode is active.
- the possibility of new instructions being loaded into the WIR while a core test is active

To meet these requirements, the data that is shifted into the WIR shift register cannot affect the active test mode of the core. Furthermore, there can be no inversion between WSI and WSO in the WIR shift register. This is to ease automation, but also to prevent failures during diagnosis.

As stated earlier, static test mode signals to the core must be controlled by the WIR. A core may come with test status terminals in addition to the test mode terminals. For instance, a test status terminal may be needed to check the status of a BIST test. To accommodate this, inputs to the WIR circuitry are allowed. Data available on these inputs terminals can be captured into the shift register upon execution of a WIR capture operation. To ease diagnosis of the captured data, the unused positions of the shift register can be filled with fixed values.

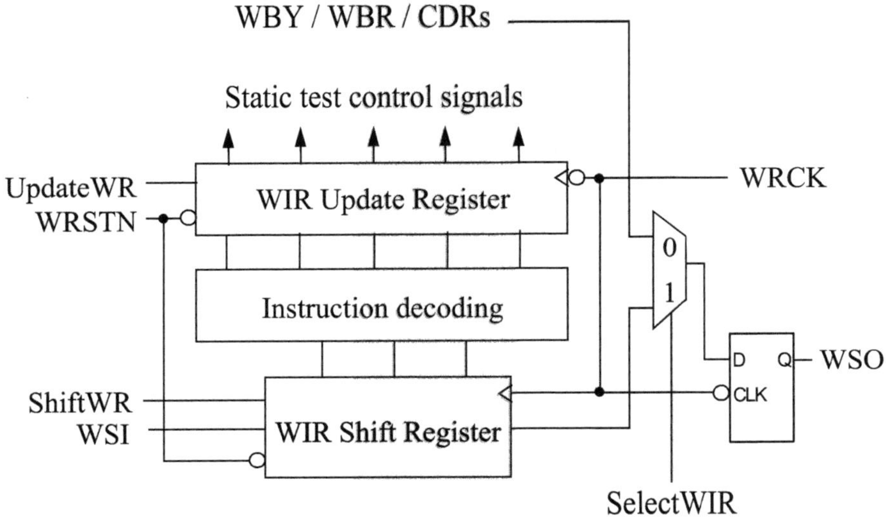

Figure 86 Example WIR design with asynchronous reset

Figure 86 shows an example WIR with an asynchronous reset implementation. The elements in the update register are resettable and output the proper control signals for functional mode (the WS_BYPASS instruction). The update register outputs do not need extra logic between them and the core since the reset put the proper values into the update register asynchronously. The shift register elements can also be made resettable such that upon reset, the shift register contains the opcode for WS_BYPASS. This serves two purposes:

- there is a known value in the shift register after reset which can be shifted out to make sure the shift register is working properly.
- if an update occurs prior to shifting an instruction into the WIR, the core will remain in functional mode.

To enable design styles that do not allow an asynchronous reset on register elements, a synchronously reset WIR implementation is allowed. The main reason that a synchronous reset may be preferred is to limit the influence of noise on the reset wiring during test. Noise on WRSTN may prevent test execution and/or diagnosis of an embedded core because the logic might reset before a test completes or before a debug action has completed. Since the reset travels all over the system chip there is a higher risk of it being affected by noise. Noise on WRSTN will not affect functional operation because the reset will result in the WIR outputting signals that puts the core in functional operation mode. An asynchronous WRSTN signal should be buffered and shielded like a clock to prevent disruption during test.

An example WIR implementation with synchronous reset of the update register is depicted in Figure 87 (page 194). Combinational logic has been added to the update register outputs to force the WS_BYPASS instruction to become effective when WRSTN is asserted. As a result, the state of the update register does not influence the active test mode or configuration of the WBR during an active WRSTN. The update register must be loaded with WS_BYPASS control values by the time WRSTN is de-asserted to prevent unexpected behavior. For this reason, combinational logic is added between the shift register and the update register as shown in Figure 87. This logic ensures that the WS_BYPASS control values are available at the parallel inputs of the update register when the wrapper is disabled. In addition, the UpdateWR signal is overruled by the inverted WRSTN signal. A synchronous reset requires an active WRCK clock to load the WS_BYPASS instruction into the update registers while the WIR circuitry decodes this test mode from the WRSTN signal.

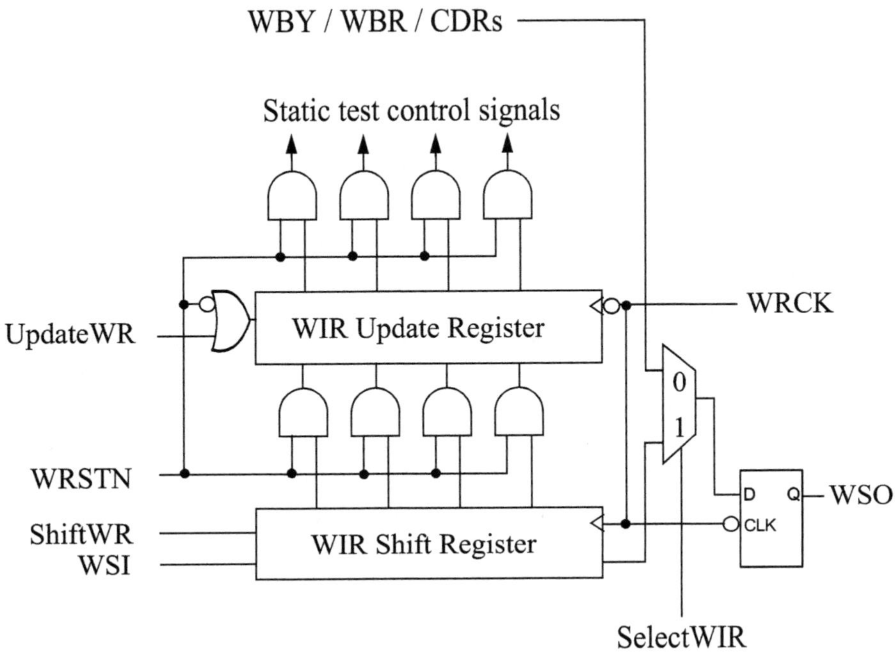

Figure 87 Example WIR Design with Synchronous Reset

1500 allows for flexibility in the design of the WIR and accompanying circuitry. The example shown in Figure 86 utilizes instruction decoding logic between the shift register and the update register. Encoded instruction opcodes result in less bits in the shift register of the WIR, preventing excessively long WIR shift registers and reducing area and static power by reducing the number of state elements. The decode logic can be implemented between the update register and the WIR outputs but this structure may allow glitches on the test control signals during instruction updates.

The WIR can also be set up without any instruction decoding logic. In this case, the shift and update registers are equal in size and there is a one-to-one correspondence between the register bit position and the test control signal created from the WIR circuitry. The main benefit of this implementation is that a complete set of combinations of control signal values can be created. If a new instruction is required, no redesign is needed as each control signal is accessible. The main drawbacks of this type of design is that many state elements may need to be added and the WIR shift chain could get excessively long. In addition, power, area and number of setup test vectors will be larger.

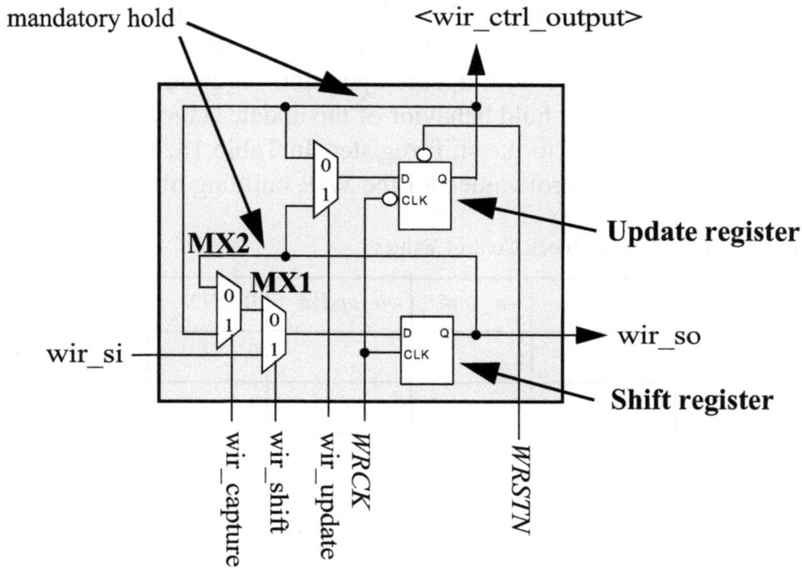

Figure 88 Example WIR Building Block

In Figure 88, an example WIR building block is shown for a WIR with no decoding. The lower portion of the figure implements a shift register segment between input serial *wir_si* and serial output *wir_so*. These signals will connect to WSI and WSO when building a complete WIR. The shift register has the capability to both shift and capture. The wir_shift and wir_capture (controlled by the mandatory ShiftWR and CaptureWR signals respectively) control which operation is occurring. If any signal other than those two signals is active, then the shift register will hold its current value. MX1 has the highest priority and enables the shift operation. MX2 chooses between the capture operation or the hold operation. Optional WIR capture inputs are typically used to observe core or wrapper internal nodes during test or debug. The example WIR building block uses the capture function to increase the test coverage of the WIR itself by observing its output. Note that the asynchronous reset of the update register that implements the normal mode can be fully tested via the capture function. The shift register changes state on the rising edge of WRCK as required by the standard. An asynchronous reset is not implemented for this register. Testing the reset of the update register in case an asynchronous reset is added to the shift register requires a protocol that differs from the standard 1149.1 Finite State Machine (FSM) protocol. This scenario is not shown here.

The upper portion of the basic building block shown in Figure 88 implements the update stage of the WIR. This update register is asynchronously reset by the active low *WRSTN* signal and changes state on the negative edge of *WRCK* in compliance with update operation requirements defined by the standard. The mandated hold behavior of the update stage is also created with a feedback path similar to the shift register. In Table 18, the WIR operations and accompanying control values for the WIR building block are listed.

Table 18 WIR Building block Control Values

WIR operation	*wir_capture*	*wir_shift*	*wir_update*	*WRSTN*	*<wir_ctrl_output>*
Functional	-	-	-	0	WS_BYPASS
Capture	1	0	0	1	static
Shift	0	1	0	1	static
Update	0	0	1	1	state change
Hold	0	0	0	1	static

8.3 Creation of the WIR for the EX example

This section describes the definition of a WIR implementation for the EX core. This process starts with identifying the control requirements during test. The following requirements are for the wrapped EX core

- Logic external to the memories will be tested with scan that makes use of a parallel access port for high data throughput and a standard scan test protocol. This protocol uses a single 'scan enable' terminal to select between the scan shift operation and scan capture operation during this test.
- During the logic test the EX core must be configured in Scan mode.
- During the MBIST test the EX core must be configured in MBIST mode.
- During interconnect test the EX core must be configured in Scan mode and its scan enable is kept inactive.
- If the wrapper is disabled or in a mode that requires functional behavior the EX core will be configured in functional mode.
- The wrapper will be able to apply static data for stuck-at test and dynamic data for sequential test, provided via the wrapper shift path.
- The WIR implementation will make use of instruction decoding.

- The WIR implementation shall make use of an asynchronous reset for the update stage of the register. This removes the requirement of a *WRCK* clock cycle before enabling the WIR.

The WIR, WBY and WBR have clear requirements for control signals. These requirements need to be addressed by the WIR circuitry. In addition, the WIR shift register must be defined.

8.3.1 Design of WIR building blocks

The standard clearly specifies the mandatory behavior of a WIR. From this, a building block that is used to construct the WIR shift register (Figure 89) and a building block that is used to construct the Update register (Figure 90), are derived. Between the shift register and update register building blocks, instruction decoding logic will be implemented for the EX core.

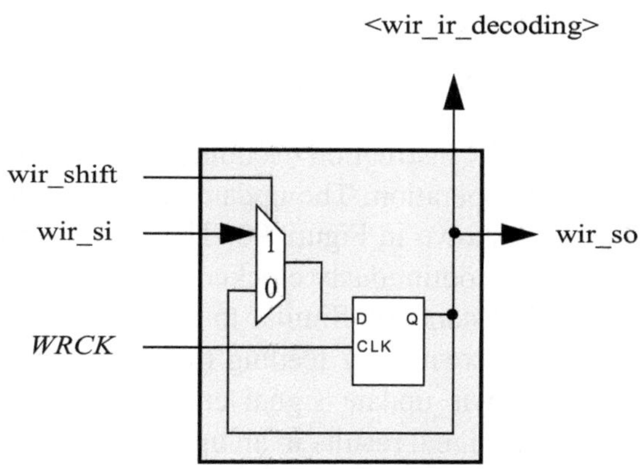

Figure 89 Building block WIR shift register

The shift register building block contains a single bit register that is continuously clocked by the wrapper clock. The WIR shift register of the EX core only needs to support the shift and hold functionality as mandated by the standard. The hold function is implemented by feeding the register output back to the register input. De-asserting the wir_shift signal enables the hold function. This signal is shared by all shift register building blocks in a WIR. Multiple building blocks are connected via the wir_si, wir_so serial data terminals to form a multi-bit shift register. The shift operation is performed by asserting the wir_shift signal. The outputs of the WIR shift register are driven into the instruction decoding logic.

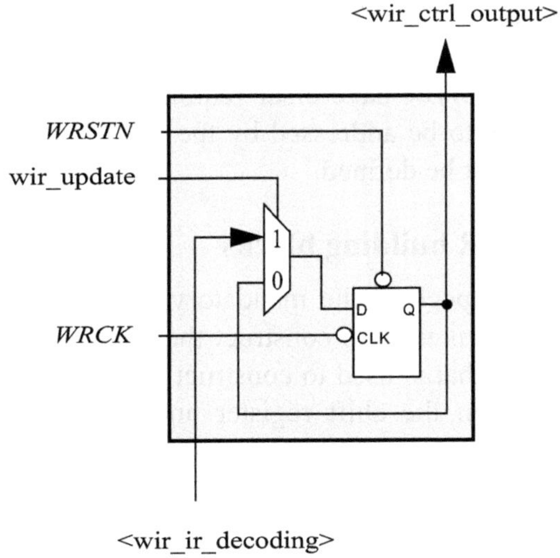

Figure 90 Building block WIR update register

The outputs of the WIR instruction decoding logic are stored in the update register after an update operation. The update register building block that is used for the EX core is shown in Figure 90. The building block consists of a single bit register that is continuously clocked by the inverted wrapper clock as mandated by the 1500 standard. Similar to the shift register building block the hold functionality is created by feeding the register output back into its input. De-asserting the wir_update signal enables the hold function while asserting the wir_update signal results in an update operation on the next falling edge of *WRCK*. The register output feeds into the wrapper or core logic to drive static control values during test. The update register of the EX core has an asynchronous reset directly driven by *WRSTN*. Asserting the *WRSTN* signal results in a reset of all update register building blocks. This forces the WS_BYPASS instruction which disables the wrapper and enables functional operation of the embedded core.

8.3.2 EX core WIR design

The WIR of the EX core wrapper implements five standard instructions and five user defined instructions. The instructions and their opcodes are listed in Table 19. This table shows that the WIR uses 4 bits in its shift register to load instruction opcodes and depicts 12 signals that are driven from the WIR update register. The EX core WIR contains four WIR shift register building blocks, instruction decoding logic, and 12 update register building blocks as depicted in Figure 91. The need for these signals and their relation to the WIR circuitry is explained in subsequent sections. The "wir_si" and "wir_so" signals feed into WSI and WSO wrapper terminals as shown in the serial port register configuration in Figure 104 (page 214). The "wir_shift" and "wir_update" signals are driven from the WIR circuitry as shown in Figure 92 (page 202). Both *WRCK* and *WRSTN* are directly driven from the standard 1500 wrapper terminals. The instruction decoding logic of the EX core is derived from Table 19. This table lists both the instruction opcodes and the accompanying output values.

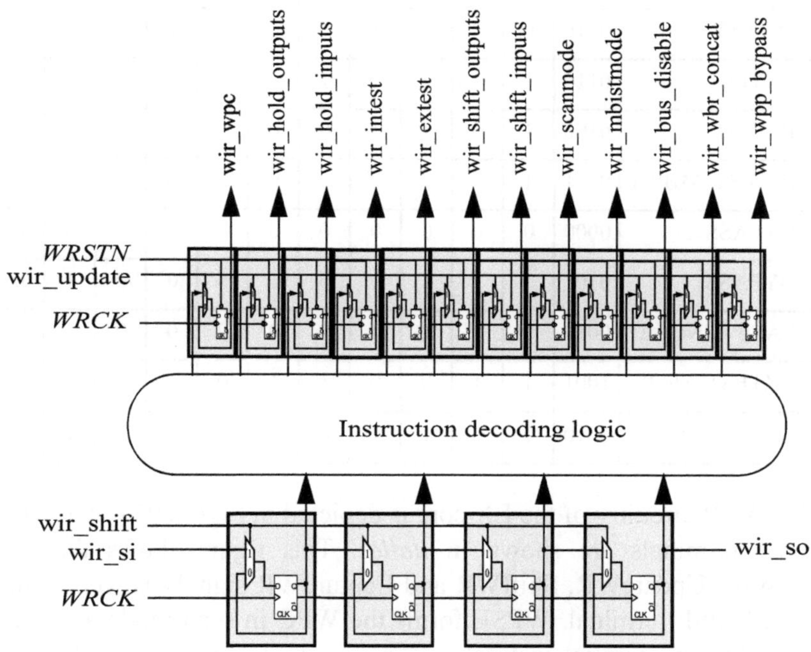

Figure 91 Implementation of the WIR for the EX core

Table 19 WIR Instruction opcodes and output values for EX wrapper

instruction	*code*	wir_wpc	wir_hold_outputs	wir_hold_inputs	wir_intest	wir_extest	wir_shift_outputs	wir_shift_inputs	wir_scanmode	wir_mbistmode	wir_bus_disable	wir_wbr_concat	wir_wpp_bypass
WS_EXTEST	0001	0	1	0	0	1	0	0	1	0	0	1	0
WS_INTEST	0010	0	0	1	1	0	0	0	1	0	1	1	0
WP_EXTEST	0110	1	1	0	0	1	0	0	1	0	0	0	0
WP_INTEST	0100	1	0	1	1	0	0	0	1	0	1	0	0
WP_INTEST_MBIST	0111	1	0	0	0	0	0	0	0	1	1	0	0
WS_BYPASS	0000	0	0	0	0	0	0	0	0	0	0	0	0
WP_BYPASS	0101	1	1	1	0	0	0	0	0	0	1	0	1
WS_SAFE_SINGLE	0011	0	1	1	0	0	0	0	0	0	1	0	0
WP_EXTEST_SEQ	1001	1	1	0	0	1	1	0	1	0	0	0	0
WP_INTEST_SEQ	1010	1	0	1	1	0	0	1	1	0	1	0	0

The WIR circuitry of the EX core is depicted in Figure 92 (page 202). The wrapper terminals are shown in *italics*. This figure shows the standard SelectWIR, UpdateWR, ShiftWR and CaptureWR signals as well as the user defined WPSE terminal. WPSE forms the WPC in scan mode and serves as scan enable for parallel instructions in this mode. As discussed in Section *4.1.2* (see page 37), the 1500 standard does not provide strict rules for the creation of the WPP. For instance, wrapper implementations that allow separate scan enables between core scan chains and wrapper scan chains are allowed by the standard. This enables scan methodologies that require different protocols between wrapper scan enable and core scan enable terminals. One application where this separation could be required is that of a pipelined scan enable methodology where a core scan enable terminal is required to transition before a wrapper scan enable terminal, ahead of capture mode becoming active in the core. For the EX core, a single scan enable terminal will be used to control both core scan chains and WBR chain segments. This allows use of standard scan protocols to test the EX core and limits requirements for integration of the EX core in an SOC.

The upper part of the logic shown in Figure 92 (page 202) controls the WIR and WBY via “wir_update”, “wir_shift”, “wir_capture” and “wby_shift” signals. These signals are independent of the signals that are driven from the WIR update register that can be found in the bottom part of the figure. In Figure 92 arrows that do not go through any gates represent the name mapping between WIR outputs and core or WBR inputs.

The WIR update register outputs feed into the WIR circuitry that implements the gating between wrapper terminals and wrapper control signals. In addition, it implements static settings for wrapper chain configuration and EX core test mode setting. Further details are discussed in subsequent sections.

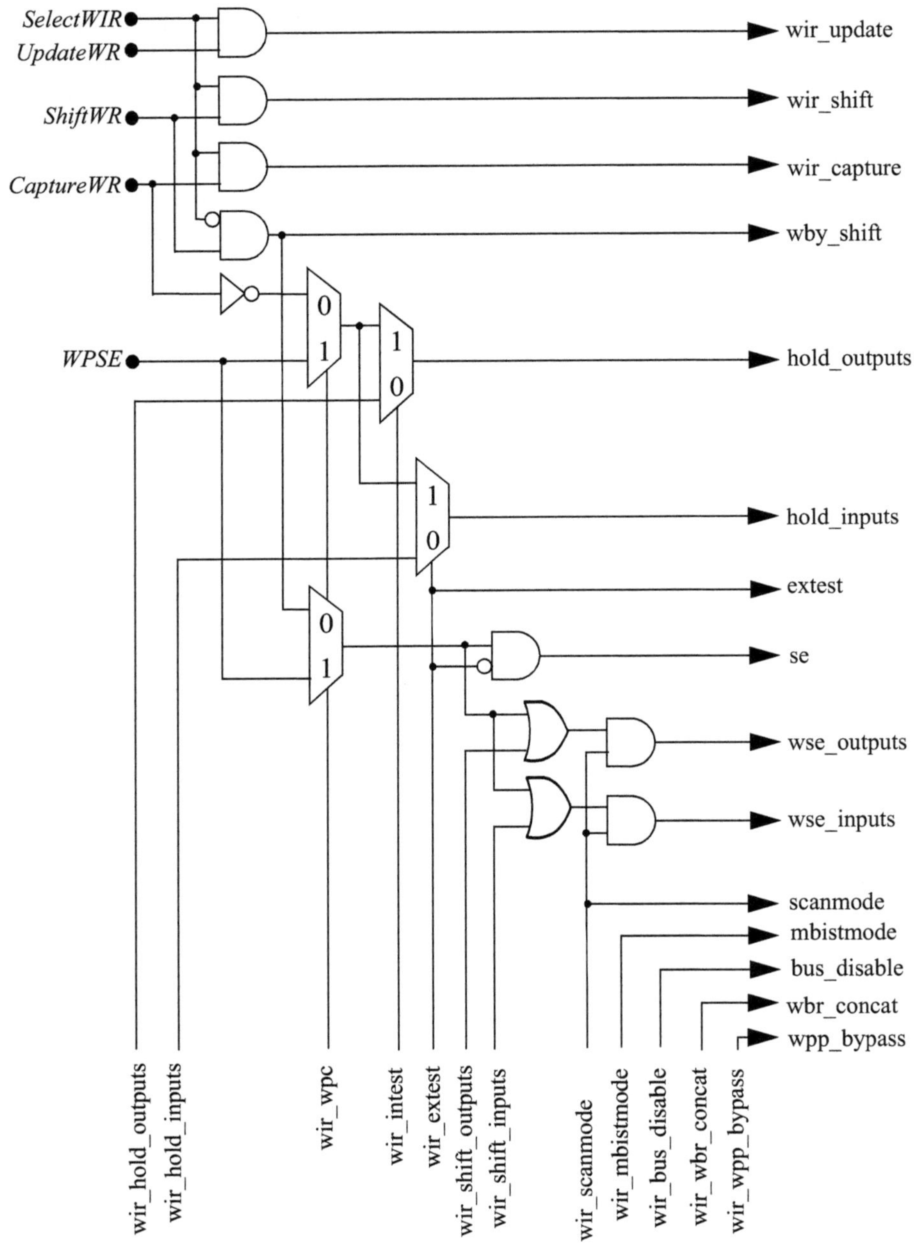

Figure 92 Implementation of the WIR circuitry for the EX core

8.3.3 Design of WIR circuitry for 1500 register control

The 1500 standard defines rules that describe expected behavior of registers mandated by the standard. Behavior of these registers is controlled by WIR circuitry that translates WIR content and primary terminal values into the correct wrapper internal control signal values. This section describes the design of the circuitry that controls the WIR, WBY and WBR.

8.3.3.1 WIR control circuitry

1500 defines that when *SelectWIR* is enabled, the WIR is operational and is controlled by the WSC signals. The circuitry that implements this behavior for the EX core WIR is shown in Figure 93. The "wir_capture", "wir_shift" and "wir_update" signals are disabled while the *SelectWIR* signal is inactive.

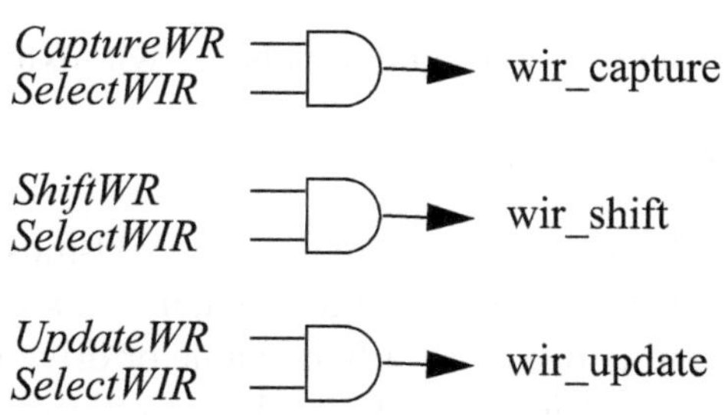

Figure 93 WIR control circuitry

8.3.3.2 WBY control circuitry

Per 1500 rules, the WBY must retain its current shift stage value whenever there are no active WBY operations, as is the case when the WIR is selected by the *SelectWIR* signal. If the WIR is not selected, the active instruction defines what register is targeted for serial access. Because the 1500 standard does not require that register content be kept upon change of active instruction, the "wby_shift" signal can be directly controlled by the *ShiftWR* signal, independent of the active instruction. This automatically implements the mandated serial port behavior requiring WBY between WSI and WSO during parallel instructions. During WP_EXTEST the WBY that is mandated between WSI-WSO follows the WBR shift operation. The final implementation of the WBY control circuitry is shown in Figure 94.

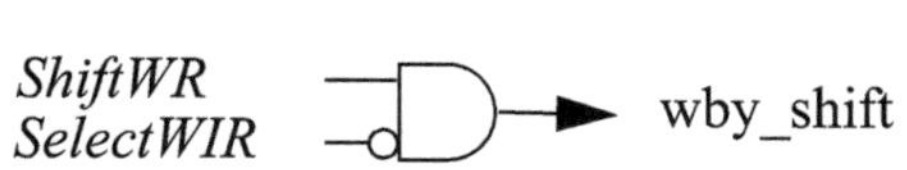

Figure 94 WBY control circuitry

8.3.3.3 WBR control circuitry

This section shows implementation of the WIR circuitry that will control the WBR. The complete set of control signals required for the EX core WBR has been defined in Section *6.6* (see page 156).

8.3.3.3.1 The "hold_inputs" signal

Generation of the "hold_inputs" signal is illustrated in Figure 95.

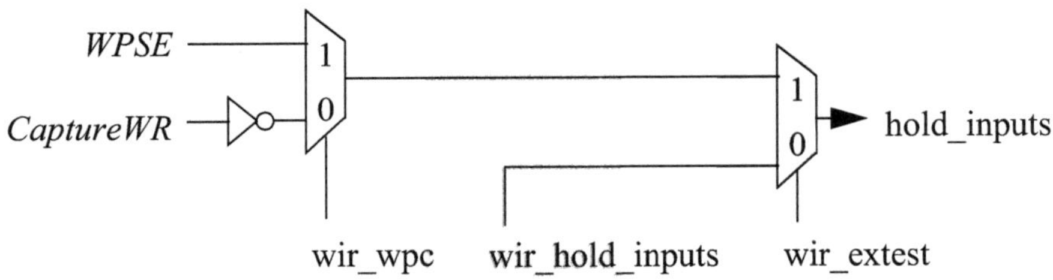

Figure 95 WBR control circuitry: hold_inputs

The behavior required of this signal is listed in the right-most column of Table 20. This table shows that the "hold_inputs" signal needs two selection mechanisms. The signal is either driven by a fixed value or driven by a wrapper terminal. In the latter case it is either driven by *~CaptureWR* (inverted *CaptureWR)* or by *WPSE.* These control signals were presented in Section *6.6.5* (see page 167).

Table 20 WIR control values for WBR control: hold_inputs

instruction	wir_wpc	wir_hold_inputs	wir_extest	hold_inputs
WS_EXTEST	0	0	1	*~CaptureWR*
WS_INTEST	0	1	0	1
WP_EXTEST	1	0	1	*WPSE*
WP_INTEST	1	1	0	1
WP_INTEST_MBIST	1	0	0	0
WS_BYPASS	0	0	0	0
WP_BYPASS	1	1	0	1
WS_SAFE_SINGLE	0	1	0	1
WP_EXTEST_SEQ	1	0	1	*WPSE*
WP_INTEST_SEQ	1	1	0	1

To control the static value and source selection, three WIR output signals have been defined. The "wir_wpc" signal selects between WPC and WSC usage. The "wir_extest" signal enables the capture function of input cells during EXTEST instructions and the "wir_hold_inputs" signal defines the fixed value of "hold_inputs" whenever instructions other than EXTEST are active.

8.3.3.3.2 The "hold_outputs" signal

Behavior of the "hold_outputs" signal is shown in the last column of Table 21 and closely resembles that of the "hold_inputs" signal. Comparison between the "hold_outputs" control circuitry shown in Figure 96 (page 206) and the "hold_inputs" circuitry in Figure 95 (page 204) shows that "hold_outputs" uses a different select signal "wir_intest" instead of "wir_extest". Naturally, this can be explained by the difference between Inward and Outward Facing modes.

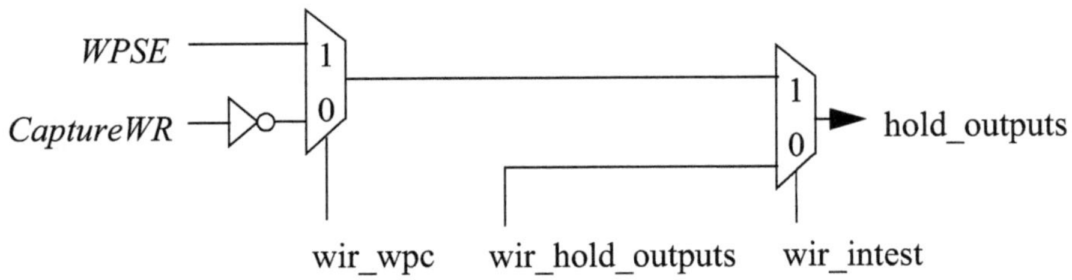

Figure 96 WBR control circuitry: hold_outputs

Table 21 WIR control values for WBR control: hold_outputs

instruction	wir_wpc	wir_hold_outputs	wir_intest	hold_outputs
WS_EXTEST	0	1	0	1
WS_INTEST	0	0	1	*~CaptureWR*
WP_EXTEST	1	1	0	1
WP_INTEST	1	0	1	*WPSE*
WP_INTEST_MBIST	1	0	0	0
WS_BYPASS	0	0	0	0
WP_BYPASS	1	1	0	1
WS_SAFE_SINGLE	0	1	0	1
WP_EXTEST_SEQ	1	1	0	1
WP_INTEST_SEQ	1	0	1	*WPSE*

8.3.3.3.3 The "wse_inputs" signal

The next two signals that are addressed are the wrapper scan enable signals wse_inputs and wse_outputs.

The "wse_inputs" signal connects to the WBR cells that isolate core input terminals. The behavior required of this signal is listed in the right-most column of Table 22. The WIR control circuitry that creates the "wse_input" signal is shown in Figure 97 (page 207). While the EX core and its wrapper are not configured in scantest mode, the "wse_inputs" signal is driven by a fixed logic '0' value. The "wir_wpc" signal selects between WPC or WSC usage. It is important to note that the *ShiftWR* signal is gated with an inverted

SelectWIR. This gating guarantees that all WBR cells hold their value during WIR operations whilst a serial instruction (e.g. WS_EXTEST) is active. The "wse_inputs" signal is driven by a fixed logic '1' value via the "wir_shift_inputs" signal during the WP_INTEST_SEQ instruction. This implements a user-defined continuous shift operation of WBR input cells in scantest mode.

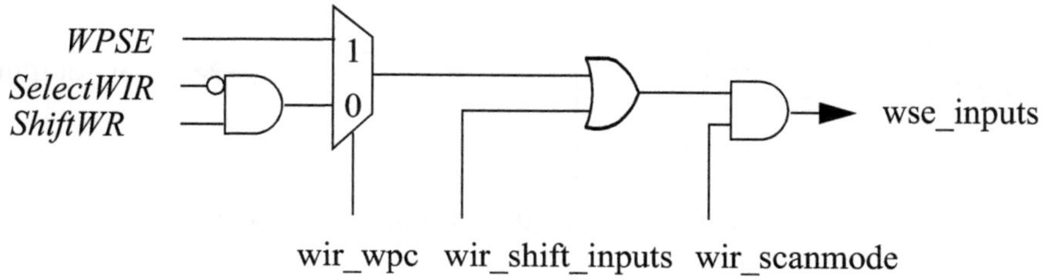

Figure 97 WBR control circuitry: wse_inputs

Table 22 WIR control values for WBR control: wse_inputs

instruction	wir_wpc	wir_shift_inputs	wir_scanmode	wse_inputs
WS_EXTEST	0	0	1	*ShiftWR*
WS_INTEST	0	0	1	*ShiftWR*
WP_EXTEST	1	0	1	*WPSE*
WP_INTEST	1	0	1	*WPSE*
WP_INTEST_MBIST	1	0	0	0
WS_BYPASS	0	0	0	0
WP_BYPASS	1	0	0	0
WS_SAFE_SINGLE	0	0	0	0
WP_EXTEST_SEQ	1	0	1	*WPSE*
WP_INTEST_SEQ	1	1	1	1

8.3.3.3.4 The "wse_outputs" signal

The behavior of the "wse_outputs" signal is shown in the last column of Table 23 and closely resembles that of the "wse_inputs" signal. Comparison between the "wse_outputs" control circuitry, shown in Figure 98, and the "wse_inputs" circuitry in Figure 97 illustrates that the only difference between these two figures is in the enabling of the continuous shift operation (with "wir_shift_inputs" for "wse_inputs" and with "wir_shift_outputs" for "wse_outputs"). The WBR output isolation cells require this continuous shift behavior during the WP_EXTEST_SEQ instruction.

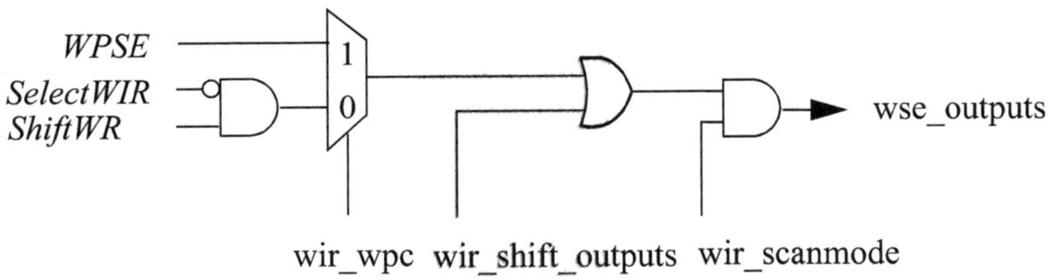

Figure 98 WBR control circuitry: wse_outputs

Table 23 WIR control values for WBR control: wse_outputs

instruction	wir_wpc	wir_shift_outputs	wir_scanmode	wse_outputs
WS_EXTEST	0	0	1	*ShiftWR*
WS_INTEST	0	0	1	*ShiftWR*
WP_EXTEST	1	0	1	*WPSE*
WP_INTEST	1	0	1	*WPSE*
WP_INTEST_MBIST	1	0	0	0
WS_BYPASS	0	0	0	0
WP_BYPASS	1	0	0	0
WS_SAFE_SINGLE	0	0	0	0
WP_EXTEST_SEQ	1	1	1	1
WP_INTEST_SEQ	1	0	1	*WPSE*

8.3.3.3.5 The "bus_disable" signal

The "bus_disable" signal is dedicated to the BC terminal of the EX core. The "bus_disable" signal is directly mapped to the "wir_bus_disable" WIR output signal as depicted in Figure 99.

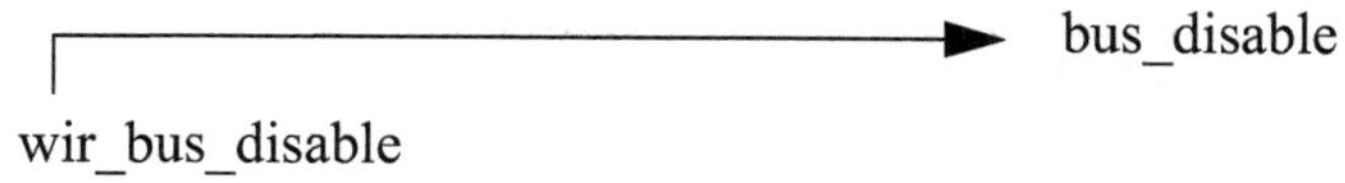

Figure 99 WBR control circuitry: bus_disable

8.3.3.4 Core control circuitry

This section presents WIR circuitry required for the control of dedicated test terminals of the EX core.

8.3.3.4.1 The scan enable signal

The scan enable signal feeds core scan chains via the "se" terminal. The WIR control circuitry that generates this signal is shown Figure 100. Values applied to the "se" terminal are explained in Table 24.

.

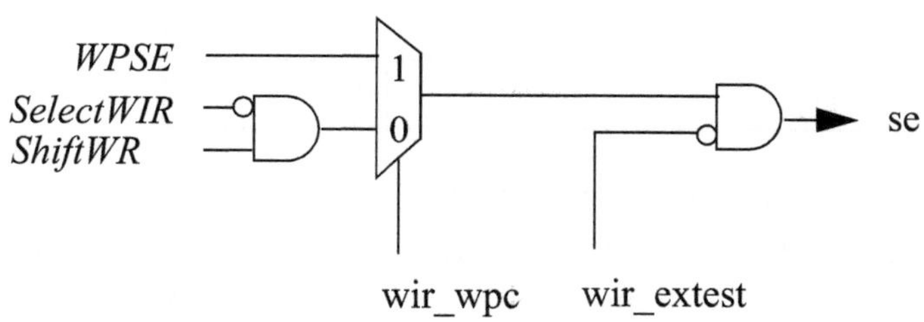

Figure 100 WBR control circuitry: se

While the EX core is not configured in scantest mode, "se" is made inactive by EX core internal logic. The "se" terminal is driven by a fixed logic '0' value during core external test modes to limit activity of core logic. The structure that selects between WPC or WSC usage is identical to the structure that is used for the "wse_inputs", and "wse_outputs" wrapper scan enable signals. A single selection mechanism is shared between these signals as shown in the EX core WIR circuitry depicted in Figure 92 (page 202).

Table 24 WIR control values for core control: se

instruction	wir_wpc	wir_extest	se
WS_EXTEST	0	1	0
WS_INTEST	0	0	*ShiftWR*
WP_EXTEST	1	1	0
WP_INTEST	1	0	*WPSE*
WP_INTEST_MBIST	-	-	-
WS_BYPASS	-	-	-
WP_BYPASS	-	-	-
WS_SAFE_SINGLE	-	-	-
WP_EXTEST_SEQ	1	1	0
WP_INTEST_SEQ	1	0	*WPSE*

8.3.3.4.2 The MBISTMODE and SCANMODE signals

MBISTMODE and SCANMODE are directly driven from the "wir_mbistmode" and "wir_scanmode" signals respectively. Figure 101 shows that these signals are directly driven by WIR outputs as required by the 1500 standard.

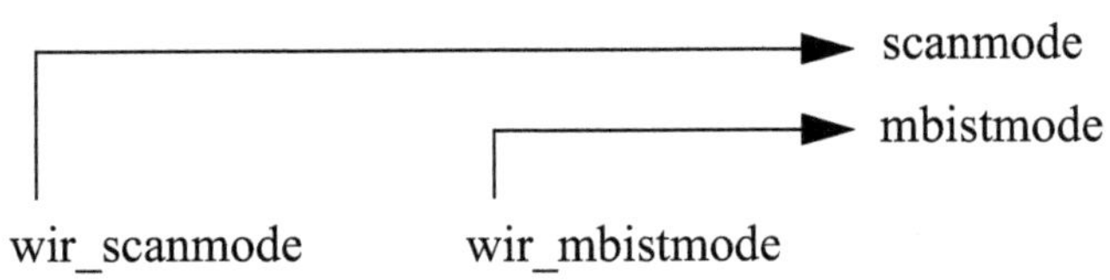

Figure 101 Core static control circuitry

The "wir_mbistmode" and "wir_scanmode" signal values corresponding to the EX core instructions are listed in Table 25.

Table 25 WIR control values for Core static control signals

instruction	wir_scanmode	wir_mbistmode	scanmode	mbistmode
WS_EXTEST	1	0	1	0
WS_INTEST	1	0	1	0
WP_EXTEST	1	0	1	0
WP_INTEST	1	0	1	0
WP_INTEST_MBIST	0	1	0	1
WS_BYPASS	0	0	0	0
WP_BYPASS	0	0	0	0
WS_SAFE_SINGLE	0	0	0	0
WP_EXTEST_SEQ	1	0	1	0
WP_INTEST_SEQ	1	0	1	0

8.3.4 Design of WIR circuitry for register configuration

So far, the WIR circuitry design for the signals that control register operation modes and test modes has been described. Another class of signals is required to control the selection of registers. This selection or configuration can be divided into two categories, the serial port register configuration and the parallel port register configuration.

8.3.4.1 Parallel port register configuration

Figure 102 shows the parallel port configuration hardware for the EX core wrapper.

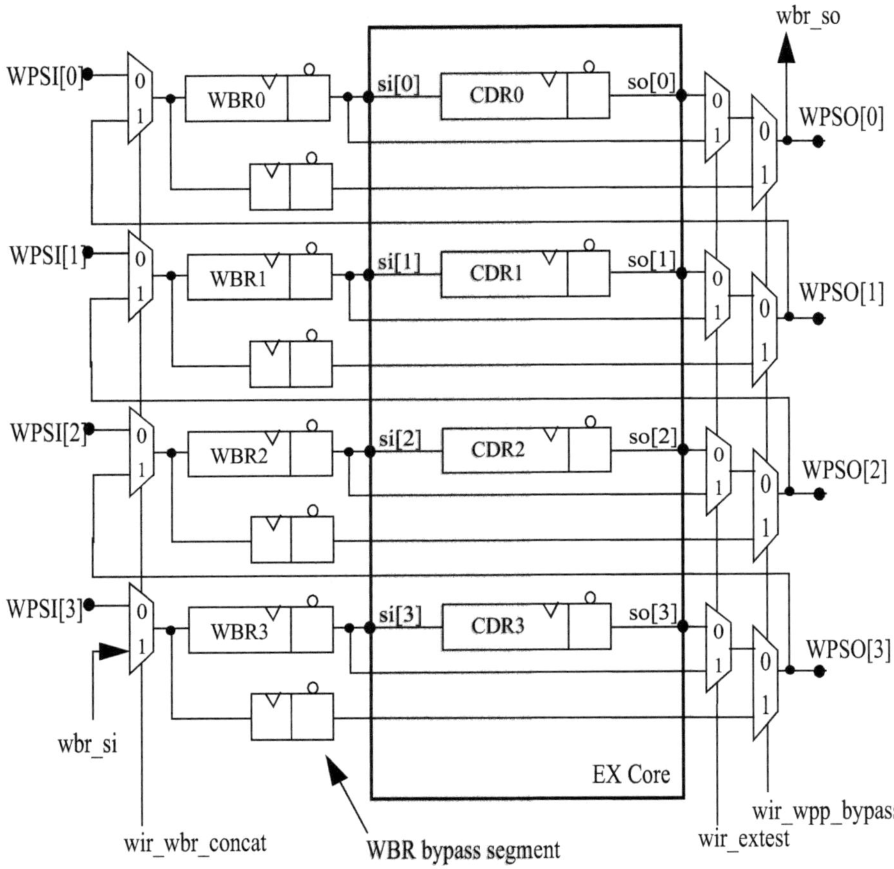

Figure 102 Parallel Port register configuration

The wrapper parallel port contains four scan input terminals WPSI[3:0] and four scan output terminals WPSO[3:0]. The number of scan input and output terminals equals the number of EX core scan chains. During internal testing of the core logic, access is needed to core scan chains as well as WBR segments. By concatenating core chains and WBR segments the full bandwidth of the parallel port is used.

For the WP_EXTEST instruction the 1500 standard mandates that only WBR segments are configured for access from WPI to WPO during the shift operation. Similarly, during the WS_EXTEST instruction, the WBR is the only register expected between WSI and WSO. To implement this behavior, a selection mechanism controlled by "wir_extest" is implemented in the wrapper. This mechanism bypasses the core internal scan chains during EXTEST instructions. The resulting configuration consists of four WBR segments between WPSI[3:0] and WPSO[3:0]. While the WS_EXTEST instruction is selected, 1500 requires that only the WBR is to be connected for serial access between WSI and WSO. To implement this behavior muxes are added between the WBR segments. These muxes, controlled by the "wir_wbr_concat" signal, concatenate all WBR segments into one chain between "wbr_si" and "wbr_so". The latter two signals will connect to WSI and WSO as shown in Figure 104 (page 214).

In addition to the registers needed for testing core internal logic and logic external to the core, a bypass path is implemented between WPSI[3:0] and WPSO [3:0]. This path (through the WBR bypass segments) is used to efficiently transport test data through the wrapper when WP_BYPASS is enabled (wir_wpp_bypass set to 1). In general, this path is used only when parallel ports of multiple cores are concatenated into one Test Access Mechanism at the SOC level. It is worth noting that all WBR bypass segments shown in Figure 102 (page 212) have a lockup latch at the chain output. These latches prevent skew problems between segments that run on different clocks and are used in different configurations.

The WBR bypass segments are clocked by the wrapper clock (WRCK). This allows the functional clock (CLK) to be switched off while WP_BYPASS is active. The WBR bypass segments hold their content in the absence of a WPP shift operation as shown in Figure 103. The complete set of control signal values for the register configuration is listed in Table 26.

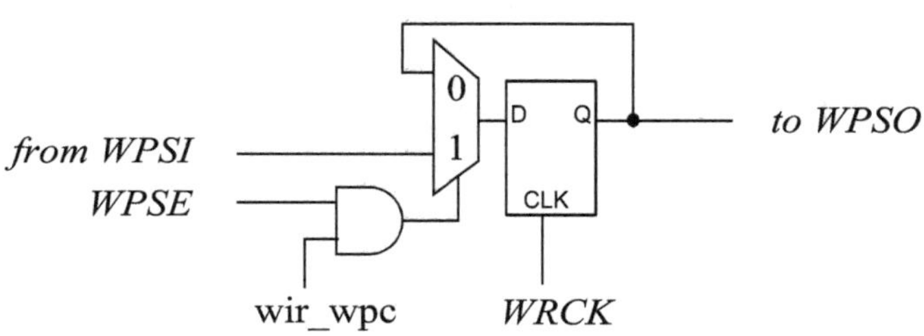

Figure 103 EX wrapper WBR bypass segment

Table 26 WIR control values for Register configuration

instruction	wir_wbr_concat	wir_extest	wir_wpp_bypass
WS_EXTEST	1	1	0
WS_INTEST	1	0	0
WP_EXTEST	0	1	0
WP_INTEST	0	0	0
WP_INTEST_MBIST	0	0	0
WS_BYPASS	0	0	0
WP_BYPASS	0	0	1
WS_SAFE_SINGLE	0	0	0
WP_EXTEST_SEQ	0	1	0
WP_INTEST_SEQ	0	0	0

8.3.4.2 Serial port register configuration

Figure 104 shows the serial port configuration hardware for the EX core wrapper.

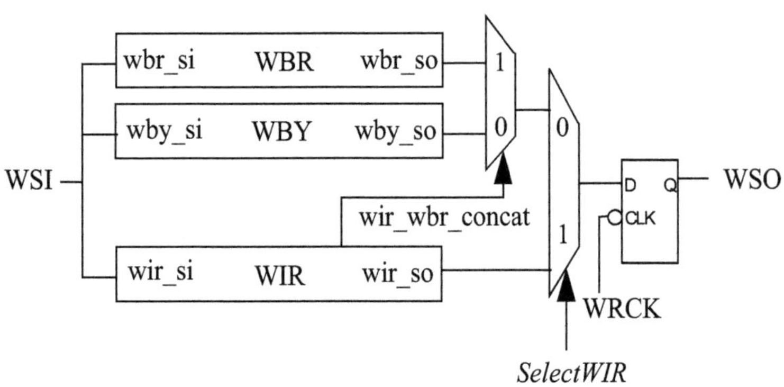

Figure 104 Serial Port register configuration

The 1500 standard mandates that, by default, the WBY is selected between WSI and WSO while the WIR is not selected. To meet this requirement, the EX core WBY will be deselected only when an instruction actively selects a different register between "wbr_si" and "wbr_so". For this purpose, the "wir_wbr_concat" signal is used to select between WBY and WBR. Note that during the WS_INTEST instructions both the WBR segments and EX core scan chains are connected between "wbr_si" and "wbr_so".

8.3.5 WIR circuitry for clock selection

The use of functional clocks is allowed in the 1500 wrapper because functional storage elements may be used in the creation of WBR cells. These clocks are referred to as auxiliary clocks when action is required on their corresponding wrapper terminal in order to execute 1500 operations in serial mode. This was discussed in Section *4.1.1.9* (see page 35). The EX wrapper implements an alternate way of handling functional clocks which makes use of bypass circuitry to allow WRCK to operate the functional storage elements during 1500 operations. This satisfies the requirement that WRCK must remain in control of sequential operation of the wrapper during serial instructions. Figure 105 shows a logical representation of the corresponding circuitry. In this figure, the 'wir_wbr_concat' signal that was introduced in Section *8.3.4.1* (see page 212) is reused to connect the EX core clock CLK to the wrapper clock WRCK. The original function of the 'wir_wbr_concat' signal which consists in configuring the WBR into single-chain mode for the WS_EXTEST and WS_INTEST instructions is maintained. The values applied to the EX core CLK terminal are explained in Table 27.

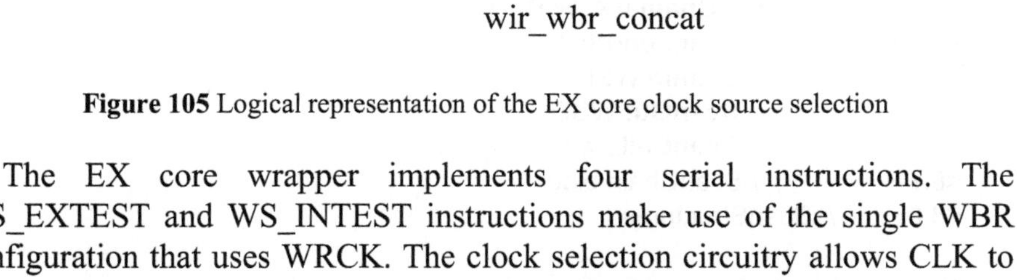

Figure 105 Logical representation of the EX core clock source selection

The EX core wrapper implements four serial instructions. The WS_EXTEST and WS_INTEST instructions make use of the single WBR configuration that uses WRCK. The clock selection circuitry allows CLK to reach the core terminal during WS_BYPASS and WS_SAFE_SINGLE. However, it should be noted that these instructions do not make use of any storage element clocked by CLK, except for the fact that WS_BYPASS enables the functional mode and makes CLK is available for this mode.

Table 27 WIR control values for Functional clock source selection

instruction	wir_wbr_concat	EX core CLK
WS_EXTEST	1	*WRCK*
WS_INTEST	1	*WRCK*
WP_EXTEST	0	*CLK*
WP_INTEST	0	*CLK*
WP_INTEST_MBIST	0	*CLK*
WS_BYPASS	0	*CLK*
WP_BYPASS	0	*CLK*
WS_SAFE_SINGLE	0	*CLK*
WP_EXTEST_SEQ	0	*CLK*
WP_INTEST_SEQ	0	*CLK*

8.4 Describing the WIR in CTL for the EX Core

From a CTL perspective the WIR is a scan chain and therefore will be defined using the **ScanStructures** syntax defined in Section *6.7.3* (see page 173). The resulting CTL description follows:

```
<8.1>   ScanStructures EX_wrapper_WSP_chains {
<8.2>       ScanChain EX_WIR_chain {
<8.3>           ScanLength 4;
<8.4>           ScanIn WSI;
<8.5>           ScanOut WSO;
<8.6>           ScanCells wir_cells[0..3];
<8.7>       } // ScanChain block
<8.8>   } // ScanStructures
```

Where:

- EX_wrapper_WSP_chains is a **ScanStructure** group containing all EX wrapper scan chains
- EX_WIR_chain is the name of the EX wrapper WIR chain
- wir_cells[0..3] represents the WIR shift stage register names

Chapter 9
Hierarchical Cores

9.1 Dealing with Hierarchical Cores

1500 can be utilized on hierarchical cores. A hierarchical core contains another full core within it labeled as child core (see Figure 106).

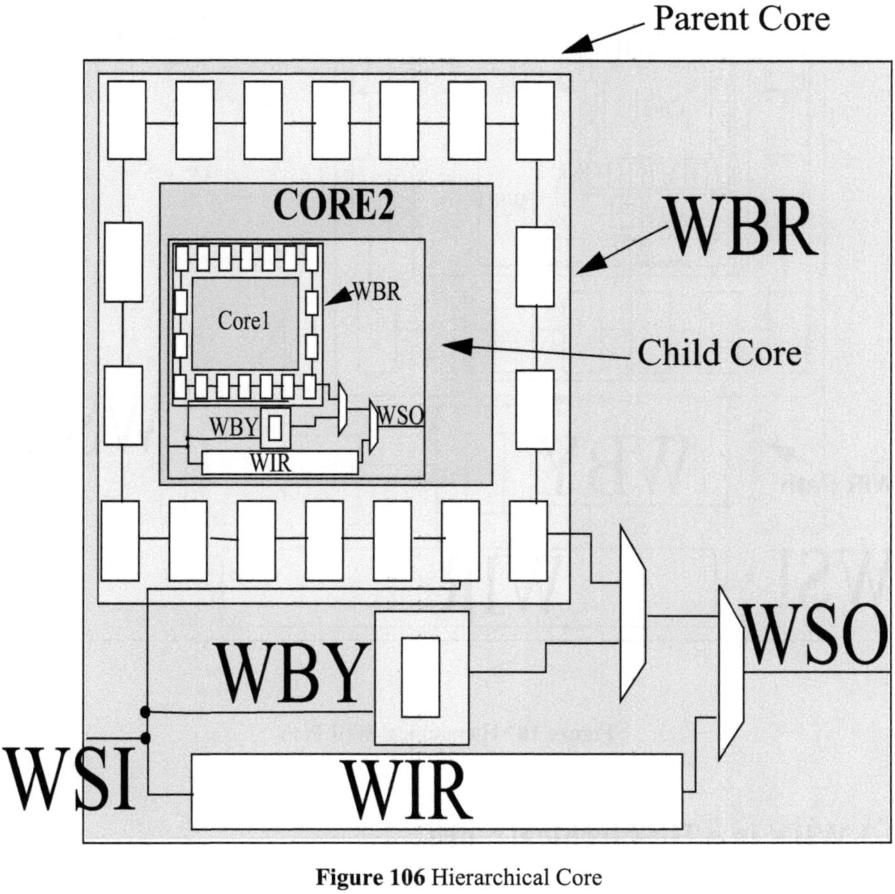

Figure 106 Hierarchical Core

Core1 is the child core and is contained within Core2, the parent core.

9.2 WIR in a Hierarchical Core

In the case of a hierarchical core, each core's WIR must be controllable through the WSP. One way to do this is to concatenate the WIR of Core1 with the WIR of Core2 and control them through the WSP as shown in Figure 107. Only the WSI is shown in Figure 107; the remaining signals of the WSP are implied.

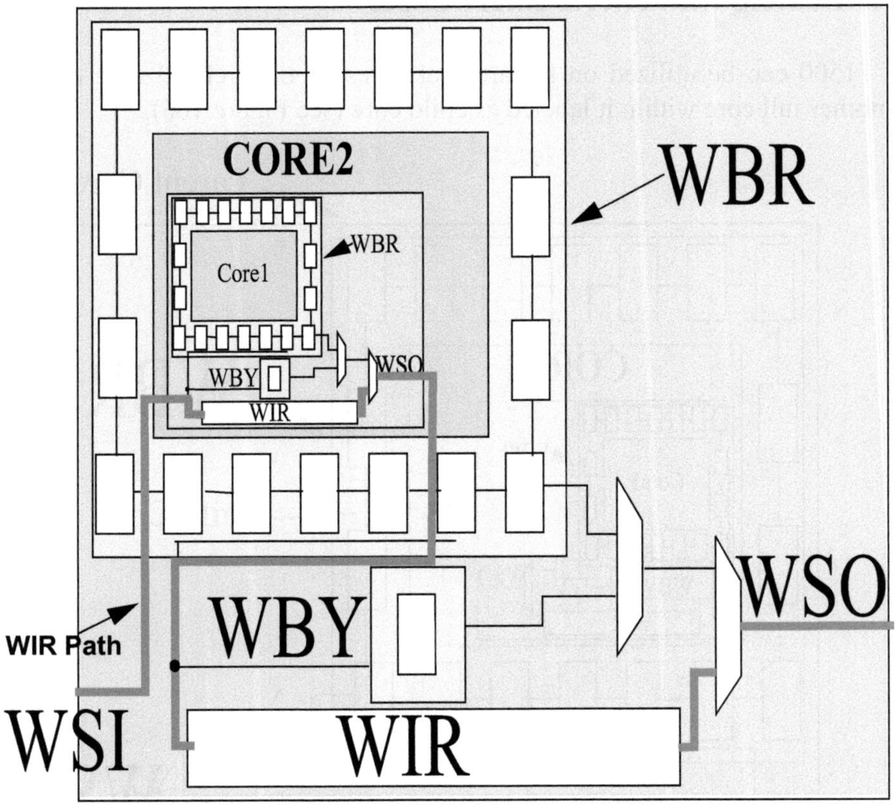

Figure 107 Hierarchical WIR Path

9.3 WBY in a Hierarchical Core

There are two possible ways that the WBY path can be set up in a hierarchical core. The first way is shown in Figure 108 where only the WBY of the parent core is utilized during the WS_BYPASS instruction or any other instruction that utilizes the WBY path.

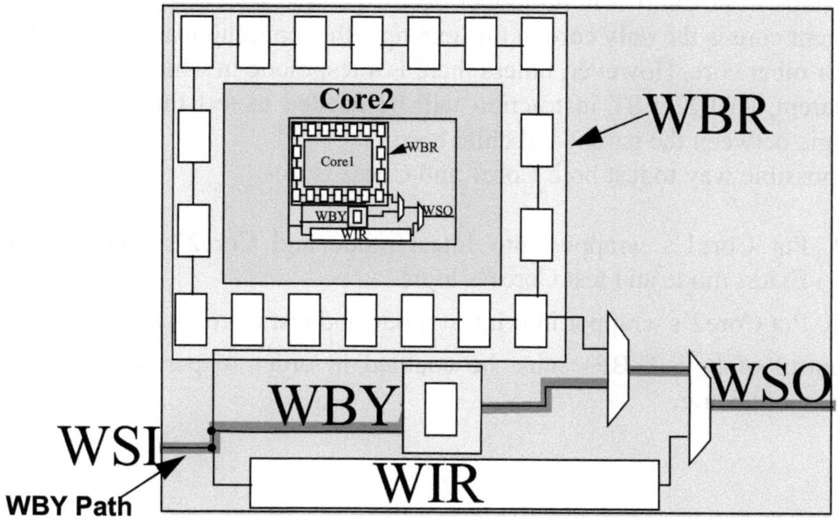

Figure 108 Embedded 1500 Core with non-hierarchical WBY path

The second way, shown in Figure 109, concatenates Core1's WBY with Core2's WBY. These two WBYs would then be connected to the SOC's WDR path.

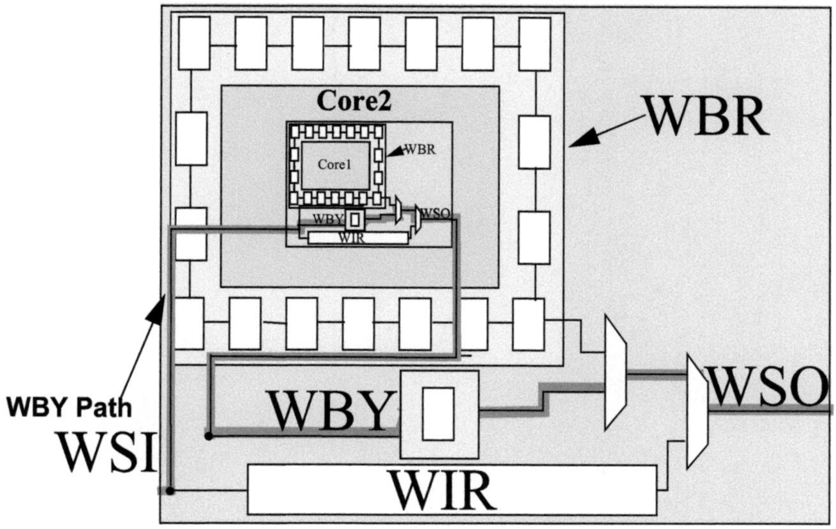

Figure 109 Embedded 1500 Core with Hierarchical WBY path

9.4 WBR in a Hierarchical Core

In a hierarchical core configuration, the WBR of the child core must be controllable. The child core no longer needs the WS_EXTEST instruction as the parent core is the only core in this example that communicates to the SOC logic or other core. However, unless there is a test mode in which the WBR is transparent, an EXTEST instruction will be needed to test the interconnect and logic between the parent and child core.

A possible way to test both Core1 and Core2 is to:

1. Put Core1's wrapper into Intest mode and Core2's wrapper into Extest mode and test Core1's logic
2. Put Core2's wrapper into Intest mode and test Core2's logic

Control of both WBRs must be enabled in order to permit the above sequence of testing.

Chapter 10
Finalizing the Wrapper Solution for the EX Core

This chapter will combine the information from the previous chapters and complete the build of the 1500 wrapper for the EX core.

10.1 Expressing Compliance to the 1500 Standard

A primary condition to 1500 compliance is that the various 1500 wrapper components must be described in CTL as shown throughout this book. However, the CTL language syntax is expected to evolve and this may invalidate CTL compliance keywords required by 1500. For this reason, compliance to 1500 is attached to a specific version of the CTL language which is the 2005 version of 1450.6.

As discussed in Section *1.2* (see page 7), there are two types of wrapper compliance to the 1500 standard. The type of compliance achieved by a core or wrapped core is to be specified in the accompanying CTL through use of the CTL **Compliancy** keyword. The corresponding CTL syntax follows:

```
<10.1>   Environment {
<10.2>          CTLMode {
<10.3>             Compliancy < IEEE1500 EXT_VERSION <Wrapped |
                   Unwrapped>| None | User USER_DEFINED > ;
<10.4>          } // CTLMode block
<10.5>   } // Environment block
```

For 1500 specific usage, the above syntax can be simplified to the following:

```
<10.6>   Environment {
<10.7>          CTLMode {
<10.8>   Compliancy < IEEE1500 EXT_VERSION <Wrapped | Unwrapped>> ;
<10.9>          } // CTLMode block
<10.10>   } // Environment block
```

The above simplified syntax eliminates the need for invalid 1500 usage of the more verbose CTL syntax. In particular, the case where no compliance is provided (**Compliancy None**) and the case where a user-defined compliance is provided (**Compliancy User** user_defined) were eliminated from the simplified and 1500-specific syntax.

1450.6 defines that a default value of "**None**" is to be affected to the **Compliancy** keyword and therefore the absence of this keyword from the CTL represents the default condition of no compliance specific features in the CTL code provided.

10.1.1 Expressing 1500 compliance for the EX wrapper

The following CTL code expresses 1500 compliance for the EX wrapper.

```
<10.11>   Environment EX_WRAPPER {
<10.12>           CTLMode {
<10.13>       Compliancy IEEE1500 2005 Wrapped;
<10.14>           } // CTLMode block
<10.15>   } // Environment block
```

10.2 The Wrapped Core

10.2.1 Wrapping the EX Core

Section *3.2* (see page 21) describes the signals on the unwrapped EX core. Section *6.5* (see page 151) describes the WBR for the EX core. Section *7.4* (see page 185) describes the WBY for the EX core and Section *8.3* (see page 196) describes the WIR for the EX core. What has not been shown are the connections between each of these entities. Figure 110 (page 223) shows the connections between each of the 1500 entities and the EX Core.

Note that most of the control signals come out of the WIR and go to the WBR and WBY. These signals are controlled by the various instructions that are shifted into the WIR.

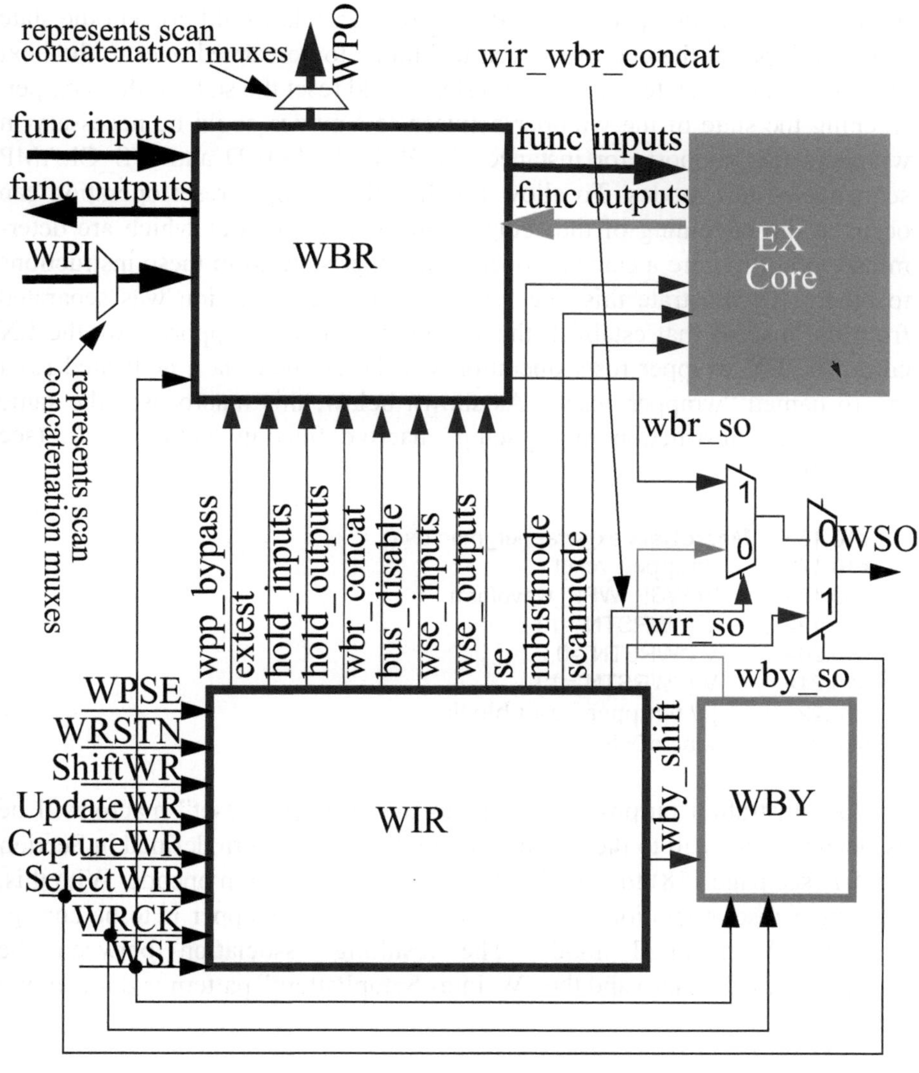

Figure 110 1500 Component Connections

10.3 Defining the wrapper reset operation in CTL

The reset operation initializes the wrapper to predetermined storage element states which ensure predetermined behavior in the wrapper. This is an essential step in the execution of tests in the wrapped core. One way to initiate a wrapper reset could be to include a reset operation at the beginning of the macro that establishes a given test mode. This way, the wrapper will be reset before the test mode opcode is loaded. However, this would prevent the state of the wrapper at the end of a given test mode to be carried over to the next test mode since each test mode setup step would reset the state of the wrapper. Keeping the state of the wrapper between test modes could be necessary in wrappers that support, for instance, the WS_PRELOAD and WS_CLAMP sequence of test modes. To allow for this, the wrapper reset operation can occur at the beginning of the setup of specific test modes which are determined to not require a carrying over of wrapper state from these instructions to others. To illustrate this, the EX wrapper reset operation was separated from the macros that establish the various test modes supported by the EX core. The EX wrapper reset operation will be defined via a dedicated reset macro named "wrapper_reset". As shown below, this macro uses the same syntax as the "instruction_mode_setup" macro defined in Section *5.5.1.3* (see page 109).

```
<10.16>  MacroDefs ex_wrapper_macros {
<10.17>     wrapper_reset {
<10.18>     W WSP_WPP_waveform;
<10.19>     V { "WRSTN"=1;}
<10.20>     V {"WRSTN"=0;}
<10.21>     V { "WRSTN"=1;}
<10.22>     } // wrapper_reset block
<10.23>  } // MacroDefs block
```

For illustration purposes, the call to the "wrapper_reset" macro will be associated to the call to the "WsIntestSetupPattern" pattern defined in Section *5.5.5.1* (see page 118) to load the WS_INTEST instruction opcode. With this, a wrapper reset operation will occur only when the wrapper is to be configured in WS_INTEST mode. The resulting association between the "wrapper_reset" macro and the "WsIntestSetupPattern" pattern is done in two steps:

1. By declaring the "wrapper_reset" macro as a setup macro in the **Protocol** statement of the "WsIntestSetupPattern" Setup macros were introduced in Section *5.3.4.1* (see page 100). The resulting CTL code follows:

```
<10.24>     PatternInformation {
<10.25>         Pattern WsIntestSetupPattern {
<10.26>             Purpose EstablishMode;
<10.27>             Protocol Macro instruction_mode_setup wrapper_reset;
<10.28>         } // Pattern block
<10.29>     } // PatternInformation block
```

2. By adding the **Setup** construct – defined in Section *5.3.1.1* (see page 93) – to the definition of the "WsIntestSetupPattern". CTL defines that this construct uses the setup macro identified above with the **Protocol** statement. The resulting "WsIntestSetupPattern" follows:

```
<10.30>   Pattern WsIntestSetupPattern {
<10.31>   Setup { }
<10.32>       P {"WSI"=0010;}
<10.33>   }
```

This means that the "wrapper_reset" macro will be invoked because of the presence of the **Setup** keyword and because a setup macro was declared in the **Protocol** statement defined by the **PatternInformation** block. There is no need to pass an argument through this call to the "wrapper_reset" macro because the macro does not need any additional information to run. This setup call will occur, as defined by CTL, at the initial stage (or setup stage) preceding the application of the WS_INTEST opcode (0010) to the wrapper.

10.4 Pattern examples

10.4.1 WS_INTEST pattern example

An example pattern is shown in Figure 111 (page 226). The pattern loads the WS_INTEST instruction into the WIR. Then begins to load data into the WBR.

Core Pattern

```
SSUCWWWW
EHPARRSS
LIDPSCIO
EFATTK
CTTUN
TWER
WRWE
I RW
R  R

1000110X
1000010X
1000110X
1100110X
1100110X
1100111X
1100110X
1010110X
1000110X
0000110X
--------
0100111X
0100110X
0100110X
0100111X
0100111X
0100110X
0100111X
```

Figure 111 WS_INTEST patterns

For convenience, the WS_INTEST patterns shown in Figure 111 include the loading of the WS_INTEST instruction opcode into the WIR as well as the first seven cycles of the execution of this instruction. The two steps are separated by a line in this figure. The CTL code corresponding to the opcode loading step was defined in Section *10.3* (see page 223). The CTL code corresponding to the latter part of the pattern set will be defined in this section. The process for creating this CTL code will consist of defining a macro, or protocol, and the pattern data. This process is identical to that discussed in Section *5.5.3* (see page 113).

The protocol corresponding to the beginning of the example WS_INTEST test will essentially consist in loading initial test data into the WBR. The corresponding CTL macro (called run_ws_intest) follows.

```
MacroDefs ex_wrapper_macros {

    wrapper_reset {
    // Refer to Section 10.3 (see page 223)
    } // wrapper_reset block

    instruction_mode_setup {
    // Refer to Section 5.5.1.3 (see page 109)
    } // instruction_mode_setup block

    run_ws_intest {
    W WSP_WPP_waveform;
    C {"ShiftWR"=1; "CaptureWR"=0; "SelectWIR"=0; "UpdateWR"=0;
        "WRSTN"=1; }
    Shift {
        V { "WSI"=#; "WSO"=#; "WRCK"=P;}
    }
    V {"ShiftWR"=0; "CaptureWR"=1; "WRCK"=P;}
    } // run_ws_intest block
} // ex_wrapper_macros block
```

The pattern data matching the above protocol can be extracted from Figure 111 (page 226) as the following:

```
Pattern WsIntestRunPattern {
    P {"WSI"=1; "WSO"=X;}
    P {"WSI"=0; "WSO"=X;}
    P {"WSI"=0; "WSO"=X;}
    P {"WSI"=1; "WSO"=X;}
    P {"WSI"=1; "WSO"=X;}
    P {"WSI"=0; "WSO"=X;}
    P {"WSI"=1; "WSO"=X;}
} // WsIntestRunPattern block
```

10.4.2 WP_INTEST pattern example

Figure 112 (page 228) shows an example pattern that loads the WP_INTEST instruction into the WIR and then begins loading the WBR.

Core Pattern

```
C S S U C W W W W W W W W W W W W W W
L E H P A R R S S P P P P P P P P P
K L I D P S C I O S S S S S S S S S
  E F A T T K     E I I I I O O O O
  C T T U N         3 2 1 0 3 2 1 0
  T W E R
  W R W E
  I   R W
  R     R

11000110X00000XXXX
11100110X00000XXXX
11100111X00000XXXX
11100110X00000XXXX
11100110X00000XXXX
11010110X00000XXXX
11000110X00000XXXX
10000110X00000XXXX
-------------------
1XXXX11XX10000XXXX
1XXXX11XX10010XXXX
1XXXX11XX11100XXXX
1XXXX11XX10000XXXX
1XXXX11XX10110XXXX
1XXXX11XX10000XXXX
1XXXX11XX11100XXXX
```

Figure 112 WP_INTEST patterns

Figure 112 shows a line separating the WP_INTEST instruction loading step from the actual test as was done for the WS_INTEST instruction in Figure 111. The macro corresponding to the instruction run follows.

```
<10.62>  MacroDefs ex_wrapper_macros {
<10.63>
<10.64>      wrapper_reset {
<10.65>      // Refer to Section 10.3 (see page 223)
<10.66>      } // wrapper_reset block
<10.67>
<10.68>      instruction_mode_setup {
<10.69>      // Refer to Section 5.5.1.3 (see page 109)
<10.70>      } // instruction_mode_setup block
<10.71>
<10.72>      run_ws_intest {
<10.73>  // Refer to Section 10.4.1 (see page 225)
<10.74>      } // run_ws_intest block
```

```
<10.75>    run_wp_intest {
<10.76>    W WSP_WPP_waveform;
<10.77>    C {"WPSE"=1; "WRSTN"=1; "SelectWIR"=X; "ShiftWR"=X;
           "UpdateWR"=X; "CaptureWR"=X; "WSI"=X; "WSO"=X;}
<10.78>    Shift {
<10.79>        V { "WPSI[0..3]"=####; "WPSO[0..3]"=####;
               "WRCK"=P; "CLK"=P;}
<10.80>    }
<10.81>    V {"WPSE"=0; "CLK"=P; "WRCK"=P;}
<10.82>    } // run_wp_intest block
<10.83>  } // ex_wrapper_macros block
```

The pattern corresponding to Figure 112 (page 228) is the following:

```
<10.84>  Pattern WpIntestRunPattern {
<10.85>      P {"WPSI[0..3]"=0000; "WPSO[0..3]"=XXXX;}
<10.86>      P {"WPSI[0..3]"=0100; "WPSO[0..3]"=XXXX;}
<10.87>      P {"WPSI[0..3]"=0011; "WPSO[0..3]"=XXXX;}
<10.88>      P {"WPSI[0..3]"=0000; "WPSO[0..3]"=XXXX;}
<10.89>      P {"WPSI[0..3]"=0110; "WPSO[0..3]"=XXXX;}
<10.90>      P {"WPSI[0..3]"=0000; "WPSO[0..3]"=XXXX;}
<10.91>      P {"WPSI[0..3]"=0011; "WPSO[0..3]"=XXXX;}
<10.92>  } // WpIntestRunPattern block
```

10.5 Combined CTL Code for the EX Wrapper

The following partial CTL code has been defined in previous chapters:

- CTL code for EX wrapper terminals
- CTL code for EX wrapper WBR and WIR and WBY
- CTL code EX wrapper instructions and setup pattern
- CTL code for EX wrapper 1500 interface timing

A combination of these various CTL sections results in the following CTL code. In addition, the association between the macros and pattern examples described in Section *10.4* (see page 225) is done here to complete the definition of these patterns in CTL.

```
<10.93>  STIL 1.0 {
<10.94>  Design 2005;
<10.95>  CTL 2005;
<10.96>  }
```

```
Signals {
   WRCK In;
   WRSTN In;
   SelectWIR In;
   ShiftWR In;
   CaptureWR In;
   UpdateWR In;
   WSI In;
   WPSI[0..3] In;
   WPSE In;
   WSO Out;
   WPSO[0..3] Out;
   CLK In;
   RESET In;
   ADDR[0..5] In;
   DIN[0..7] In;
   DOUT[0..7] Out;
   READY In;
   BC Out;
   ACK Out;
   RX Out;
   TX Out;
   MBISTDLOG In;
   MBISTRUN In;
   MBISTDLOGOUT Out;
   MBISTDONE Out;
   MBISTFAIL Out;
}

SignalGroups {
   all_functional='CLK + RESET + ADDR[0..5] + DIN[0..7] +
       DOUT[0..7] + READY + BC + ACK + RX + TX';
   wpp_mbist='MBISTDLOG + MBISTRUN +
   MBISTDLOGOUT + MBISTDONE + MBISTFAIL + CLK';
   wpp_scan='WPSI[0..3] + WPSO[0..3] + WPSE + CLK';
   wpp_wrck='WRCK';
   wsp_wrck='WRCK';
}

MacroDefs ex_wrapper_macros {

   wrapper_reset {
      W WSP_WPP_waveform;
      V { "WRSTN"=1;}
      V {"WRSTN"=0;}
      V { "WRSTN"=1;}
   } // wrapper_reset block
```

```
<10.142>    instruction_mode_setup {
<10.143>        W WSP_WPP_waveform;
<10.144>        V {"WRCK"=P; "WRSTN"=1; "UpdateWR"=0;}
<10.145>        V {"WRSTN"=0;}
<10.146>        V {"WRSTN"=1;}
<10.147>    V {"SelectWIR"=1; }
<10.148>    V {"ShiftWR"=1; }
<10.149>    Shift {
<10.150>        V { "WSI"=#; "WRCK"=P;}
<10.151>    }
<10.152>    V { "WRCK"=P; "ShiftWR"=0;}
<10.153>    V {"UpdateWR"=1;}
<10.154>    V {"UpdateWR"=0;}
<10.155>    V {"SelectWIR"=0;}
<10.156>    } // instruction_mode_setup block
<10.157>
<10.158>    run_ws_intest {
<10.159>    W WSP_WPP_waveform;
<10.160>    C {"ShiftWR"=1; "CaptureWR"=0; "SelectWIR"=0; "UpdateWR"=0;
                "WRSTN"=1; }
<10.161>    Shift {
<10.162>        V { "WSI"=#; "WSO"=#; "WRCK"=P;}
<10.163>  }
<10.164>    } // run_ws_intest block
<10.165>
<10.166>    run_wp_intest {
<10.167>    W WSP_WPP_waveform;
<10.168>    C {"WPSE"=1; "WRSTN"=1; "SelectWIR"=X; "ShiftWR"=X;
            "UpdateWR"=X; "CaptureWR"=X; "WSI"=X; "WSO"=X;}
<10.169>    Shift {
<10.170>        V { "WPSI[0..3]"=####; "WPSO[0..3]"=####;
                "WRCK"=P; "CLK"=P;}
<10.171>  }
<10.172>    } // run_wp_intest block
<10.173>
<10.174>  } // MacroDefs block
<10.175>
<10.176>  ScanStructures EX_wrapper_WSP_chains {
<10.177>    ScanChain EX_WBY_chain {
<10.178>        ScanLength 1;
<10.179>        ScanIn WSI;
<10.180>        ScanOut WSO;
<10.181>        ScanCells wby_cell;
<10.182>    } // ScanChain EX_WBY_chain block
```

```
  ScanChain EX_WIR_chain {
      ScanLength 4;
      ScanIn WSI;
      ScanOut WSO;
      ScanCells wir_cells[0..3];
  } // ScanChain EX_WIR_chain block

  ScanChain EX_WBR_WSP_chain {
      ScanLength 33;
      ScanIn WSI;
      ScanOut WSO;
      ScanMasterClock WRCK;
      ScanCells {
WBR_READY; WBR_DIN[0..7]; WBR_RESET; WBR_ADDR[0..5];
WBR_MBISTRUN; WBR_MBISTDLOG; WBR_DOUT[0..7];WBR_TX;
WBR_RX; WBR_ACK; WBR_BC; WBR_MBISTDONE;
WBR_MBISTFAIL; WBR_MBISTDLOGOUT;
      } // ScanCells block
  } // ScanChain EX_WBR_WSP_chain block
} // ScanStructures EX_wrapper_WSP_chains

ScanStructures EX_wrapper_WPP_chains {
  ScanChain EX_WBR_WPP_chain_0 {
      ScanLength 7;
      ScanIn WPSI[0];
      ScanOut WPSO[0];
      ScanMasterClock WRCK; CLK;
      ScanCells {
WBR_TX; WBR_RX; WBR_ACK; WBR_BC; WBR_MBISTDONE;
WBR_MBISTFAIL; WBR_MBISTDLOGOUT;
      } // ScanCells block
  } // ScanChain EX_WBR_WPP_chain_0 block

  ScanChain EX_WBR_WPP_chain_1 {
      ScanLength 8;
      ScanIn WPSI[1];
      ScanOut WPSO[1];
      ScanMasterClock WRCK; CLK;
      ScanCells {
          WBR_DOUT[0..7];
      } // ScanCells block
  } // ScanChain EX_WBR_WPP_chain_1 block
```

```
  ScanChain EX_WBR_WPP_chain_2 {
    ScanLength 9;
    ScanIn WPSI[2];
    ScanOut WPSO[2];
    ScanMasterClock WRCK; CLK;
    ScanCells {
      WBR_RESET; WBR_ADDR[0..5]; WBR_MBISTRUN;
     WBR_MBISTDLOG;
    } // ScanCells block
  } // ScanChain EX_WBR_WPP_chain_2 block

  ScanChain EX_WBR_WPP_chain_3 {
    ScanLength 9;
    ScanIn WPSI[3];
    ScanOut WPSO[3];
    ScanMasterClock WRCK; CLK;
    ScanCells {
      WBR_READY; WBR_DIN[0..7];
    } // ScanCells block
  } // ScanChain EX_WBR_WPP_chain_3 block

} // ScanStructures EX_wrapper_WPP_chains block

Timing {
  WaveformTable WSP_WPP_waveform {
    Period '100ns';
    Waveforms {
    "WRSTN" { 01X { '0ns' D/U/N; }}
    "SelectWIR" { 01X { '0ns' D/U/N; }}
    "ShiftWR" { 01X { '0ns' D/U/N; }}
    "CaptureWR" { 01X { '0ns' D/U/N; }}
    "UpdateWR" { 01X { '0ns' D/U/N; }}
    'WSI+WPSI[0..3]' { 01X { '0ns' D/U/N; }}
    "WPSE" { 01X { '0ns' D/U/N; }}
    'MBISTRUN+MBISTLOG' { 01X { '0ns' D/U/N; }}
    'CLK+WRCK' { P { '0ns' D; '45ns' U; '55ns' D; }}
    'WSO+WPSO[0..3]' { { '10ns' lhXt; }}
    'MBISTFAIL+MBISTDONE+MBISTLOGOUT' { { '50ns' lhXt; }}
    } // Waveforms block
  } // WaveformTable block
}// Timing block
```

```
Environment EX_wrapper {
  CTLMode {
  Compliancy IEEE1500 2005 Wrapped;
      Internal {
          READY {
              IsConnected In {
                  StateElement Scan WBR_READY;
                  Wrapper IEEE1500 CellID WC_SD1_COI;
              }
          } // READY block
          MBISTDLOG {
              IsConnected In {
                  StateElement Scan WBR_MBISTDLOG;
                  Wrapper IEEE1500 CellID WC_SD1_COI;
              }
          } // MBISTDLOG block
          MBISTRUN{
              IsConnected In {
                  StateElement Scan WBR_MBISTRUN;
                  Wrapper IEEE1500 CellID WC_SD1_COI;
              }
          } // MBISTRUN block
          MBISTDLOGOUT {
              IsConnected In {
                  StateElement Scan WBR_MBISTDLOGOUT;
                  Wrapper IEEE1500 CellID WC_SD1_COI;
              }
          } // MBISTDLOGOUT block
          MBISTDONE {
              IsConnected In {
                  StateElement Scan WBR_MBISTDONE;
                  Wrapper IEEE1500 CellID WC_SD1_COI;
              }
          } // MBISTDONE block

          MBISTFAIL {
              IsConnected In {
                  StateElement Scan WBR_MBISTFAIL;
                  Wrapper IEEE1500 CellID WC_SD1_COI;
              }
          } // MBISTFAIL block
          ADDR[0..5] {
              IsConnected In {
                  StateElement Scan WBR_ADDR[0..5];
                  Wrapper IEEE1500 CellID WC_SF1_CII;
              }
          } // ADDR[0..5] block
```

```
DIN[0..7]{
    IsConnected In {
        StateElement Scan WBR_DIN[0..7];
        Wrapper IEEE1500 CellID WC_SF1_CII;
    }
} // DIN[0..7] block
DOUT[0..7] {
    IsConnected In {
        StateElement Scan WBR_DOUT[0..7];
        Wrapper IEEE1500 CellID WC_SF1_CII;
    }
} // DOUT[0..7] block
TX {
    IsConnected In {
        StateElement Scan WBR_TX;
        Wrapper IEEE1500 CellID WC_SF1_CII;
    }
} // TX block
ACK {
    IsConnected In {
        StateElement Scan WBR_ACK;
        Wrapper IEEE1500 CellID WC_SF1_CII;
    }
} // ACK block
RX {
    IsConnected In {
        StateElement Scan WBR_RX;
        Wrapper IEEE1500 CellID WC_SF1_CII;
    }
} // RX block
RESET {
    IsConnected In {
        StateElement Scan WBR_RESET;
        Wrapper IEEE1500 CellID WC_SD1_CII_O;
    }
} // RESET block
BC {
    IsConnected In {
        StateElement Scan WBR_BC;
        Wrapper IEEE1500 CellID WC_SD1_COI_G;
    }
} // BC block
CLK {
    IsConnected In {
        Wrapper None;
    }
} // CLK block
```

```
wsp_wrck {
    DataType TestControl ScanMasterClock;
    Wrapper IEEE1500 PinID WRCK;
    DriveRequirements {
        TimingSensitive {
        Pulse High Min <t_ckwh>;
        Pulse Low Min <t_ckwl>;
        Period Min <t_ckwh> + <t_ckwl>;
        } // TimingSensitive block
    } // DriveRequirements block
} // wsp_wrck block

wpp_wrck {
    DataType TestControl ScanMasterClock;
    Wrapper User WPP PinID WPC;
    DriveRequirements {
        TimingSensitive {
        Pulse High Min <t_ckwh>;
        Pulse Low Min <t_ckwl>;
        Period Min <t_ckwh + t_ckwl>;
        } // TimingSensitive block
    } // DriveRequirements block
} // wpp_wrck block

WRSTN {
    DataType TestControl TestWrapperControl {
        ActiveState ForceDown;
    }
    Wrapper IEEE1500 PinID WRSTN;
    DriveRequirements {
        TimingSensitive {
            Pulse Low Min <t_rstl>;
            Reference WRCK {
                SelfEdge Trailing;
                ReferenceEdge Leading;
                Setup <t_rstsu>;
            } // Reference block
        } ) // TimingSensitive block
    } ) // DriveRequirements block
} // WRSTN block
```

```
SelectWIR {
    DataType TestControl TestWrapperControl {
        ActiveState ForceUp;
    }
    Wrapper IEEE1500 PinID SelectWIR;
    DriveRequirements {
        TimingSensitive {
            Reference WRCK {
                SelfEdge LeadingTrailing;
                ReferenceEdge Leading;
                Setup <tswsu>;
                Hold <tswhd>;
            } // Reference block
        } ) // TimingSensitive block
    } ) // DriveRequirements block
} // SelectWIR block
ShiftWR {
    DataType TestControl ScanEnable;
    DataType TestControl TestWrapperControl {
        ActiveState ForceUp;
    }
    Wrapper IEEE1500 PinID ShiftWR;
    DriveRequirements {
        TimingSensitive {
            Reference WRCK {
                SelfEdge LeadingTrailing;
                ReferenceEdge Leading;
                Setup <tctlsu>;
                Hold <tctlhd>;
            } // Reference block
        } ) // TimingSensitive block
    } ) // DriveRequirements block
} // ShiftWR block
CaptureWR {
    DataType TestControl TestWrapperControl {
        ActiveState ForceUp;
    }
    Wrapper IEEE1500 PinID CaptureWR;
    DriveRequirements {
        TimingSensitive {
            Reference WRCK {
                SelfEdge LeadingTrailing;
                ReferenceEdge Leading;
                Setup <tctlsu>;
                Hold <tctlhd>;
            } // Reference block
        } ) // TimingSensitive block
    } ) // DriveRequirements block
} // CaptureWR block
```

```
UpdateWR {
    DataType TestControl TestWrapperControl {
        ActiveState ForceUp;
    }
    Wrapper IEEE1500 PinID UpdateWR;
    DriveRequirements {
        TimingSensitive {
            Reference WRCK {
                SelfEdge LeadingTrailing;
                ReferenceEdge Trailing;
                Setup <tctlsu>;
                Hold <tctlhd>;
            } // Reference block
        } ) // TimingSensitive block
    } ) // DriveRequirements block
} // UpdateWR block
CLK {
    DataType TestControl ScanMasterClock;
    Wrapper User WPP PinID WPC;
    DriveRequirements {
        TimingSensitive {
            Reference WRCK {
                SelfEdge Leading;
                ReferenceEdge Leading;
                Setup <tctlsu>;
                Hold <tctlhd>;
            } // Reference block
        } // TimingSensitive block
    } // DriveRequirements block
} // CLK block

WSI {
    DataType TestData ScanDataIn {
        ScanDataType Boundary;
    }
    Wrapper IEEE1500 PinID WSI;
    StrobeRequirements {
        TimingSensitive {
            Reference WRCK {
            SelfEdge LeadingTrailing;
            ReferenceEdge Trailing;
            EarliestTimeValid <tsov>;
            } // Reference block
        } ) // TimingSensitive block
    } ) // StrobeRequirements block
} // WSI block
```

```
WSO {
    DataType TestData ScanDataOut {
        ScanDataType Boundary;
    }
    Wrapper IEEE1500 PinID WSO;
    StrobeRequirements {
        TimingSensitive {
            Reference WRCK {
                SelfEdge LeadingTrailing;
                ReferenceEdge Trailing;
                EarliestTimeValid <tsov>;
            } // Reference block
        } ) // TimingSensitive block
    } ) // StrobeRequirements block
} // WSO block
WPSI [0..3] {
    DataType Testdata ScanDataIn {
        ScanDataType Boundary;
    }
    Wrapper User WPP PinID WPI;
    DriveRequirements {
        TimingSensitive {
            Reference WRCK {
                SelfEdge Leading;
                ReferenceEdge Leading;
                Setup <tsisu>;
                Hold <tsihd>;
            } // Reference block
        } ) // TimingSensitive block
    } ) // DriveRequirements block
} // WPSI [0..3] block
WPSO[0..3] {
    DataType TestData ScanDataOut {
        ScanDataType Boundary;
    }
    Wrapper User WPP PinID WPO;
    StrobeRequirements {
        TimingSensitive {
            Reference WRCK {
                SelfEdge LeadingTrailing;
                ReferenceEdge Trailing;
                EarliestTimeValid <tsov>;
            } // Reference block
        } ) // TimingSensitive block
    } ) // StrobeRequirements block
} // WPSO[0..3] block
```

```
MBISTFAIL {
    DataType TestData ScanDataOut {
        ScanDataType Boundary;
    }
    Wrapper User WPP PinID WPO;
    StrobeRequirements {
        TimingSensitive {
            Reference CLK {
                SelfEdge LeadingTrailing;
                ReferenceEdge Leading;
                EarliestTimeValid <tsov>;
            } // Reference block
        } // TimingSensitive block
    } // StrobeRequirements block
} // MBISTFAIL block
MBISTDONE {
    DataType TestData ScanDataOut {
        ScanDataType Boundary;
    }
    Wrapper User WPP PinID WPO;
    StrobeRequirements {
        TimingSensitive {
            Reference CLK {
                SelfEdge LeadingTrailing;
                ReferenceEdge Leading;
                EarliestTimeValid <tsov>;
            } // Reference block
        } // TimingSensitive block
    } // StrobeRequirements block
} // MBISTDONE block
MBISTLOGOUT {
    DataType TestData ScanDataOut {
        ScanDataType Boundary;
    }
    Wrapper User WPP PinID WPO;
    StrobeRequirements {
        TimingSensitive {
            Reference CLK {
                SelfEdge LeadingTrailing;
                ReferenceEdge Leading;
                EarliestTimeValid <tsov>;
            } // Reference block
        } // TimingSensitive block
    } // StrobeRequirements block
} // MBISTLOGOUT block
```

```
        WPSE {
            DataType TestControl TestWrapperControl;
            DataType TestControl ScanEnable{
                ActiveState ForceValid;
            }
            Wrapper User WPP PinID WPC;
            DriveRequirements {
                TimingSensitive {
                    Reference WRCK {
                        SelfEdge LeadingTrailing;
                        ReferenceEdge Leading;
                        Setup <tctlsu>;
                        Hold <tctlhd>;
                    } // Reference block
                } // TimingSensitive block
            } // DriveRequirements block
        } // WPSE block
        MBISTRUN {
            DataType TestControl TestWrapperControl;
            Wrapper User WPP PinID WPC;
            DriveRequirements {
                TimingSensitive {
                    Reference CLK {
                        SelfEdge LeadingTrailing;
                        ReferenceEdge Leading;
                        Setup <tctlsu>;
                        Hold <tctlhd>;
                    } // Reference block
            } // DriveRequirements block
        } // MBISTRUN block
        MBISTLOG {
            DataType TestControl TestWrapperControl;
            Wrapper User WPP PinID WPC;
            DriveRequirements {
                TimingSensitive {
                    Reference CLK {
                        SelfEdge LeadingTrailing;
                        ReferenceEdge Leading;
                        Setup <tctlsu>;
                        Hold <tctlhd>;
                    } // Reference block
                } // TimingSensitive block
            } // DriveRequirements block
        } // MBISTLOG block
    } // Internal block
} // CTLMode block
```

```
CTLMode WS_BYPASS_MODE {
  TestMode Bypass Normal;
  DomainReferences {
      MacroDefs ex_wrapper_macros;
      ScanStructures EX_wrapper_WSP_chains;
  } // DomainReferences block
  Internal {
      all_functional {DataType Functional;}
      wpp_mbist {DataType Unused;}
  } // Internal block
  PatternInformation {
      Pattern WsBypassSetupPattern {
          Purpose EstablishMode;
          Protocol Macro instruction_mode_setup;
      } // Pattern block
      Macro instruction_mode_setup {
          Purpose ModeControl;
          UseByPattern EstablishMode;
      } // Macro block
  } // PatternInformation block
  TestModeForWrapper WS_BYPASS 0000;
} // CTLMode WS_BYPASS_MODE block

CTLMode WP_BYPASS_MODE {
  TestMode InternalTest;
  DomainReferences {
      MacroDefs ex_wrapper_macros;
      ScanStructures EX_wrapper_WSP_chains
                     EX_wrapper_WPP_chains;
  } // DomainReferences block
  Internal {
      all_functional {DataType Functional;}
      wpp_mbist {DataType Unused;}
  } // Internal block
  PatternInformation {
      Pattern WpBypassSetupPattern {
          Purpose EstablishMode;
          Protocol Macro instruction_mode_setup;
      } // Pattern block
      Macro instruction_mode_setup {
          Purpose ModeControl;
          UseByPattern EstablishMode;
      } // Macro block
  } // PatternInformation block
  TestModeForWrapper WP_BYPASS 0101;
} // CTLMode WP_BYPASS_MODE block
```

```
CTLMode WS_EXTEST_MODE {
  TestMode InternalTest;
  DomainReferences {
      MacroDefs ex_wrapper_macros;
  ScanStructures EX_wrapper_WSP_chains;
  } // DomainReferences block
  Internal {
      all_functional {DataType Functional;}
      wpp_mbist {DataType Unused;}
  } // Internal block
  PatternInformation {
      Pattern WsExtestSetupPattern {
          Purpose EstablishMode;
          Protocol Macro instruction_mode_setup;
      } // Pattern block

      Macro instruction_mode_setup {
          Purpose ModeControl;
          UseByPattern EstablishMode;
      } // Macro block
  } // PatternInformation block
  TestModeForWrapper WS_EXTEST 0001;
} // CTLMode WS_EXTEST_MODE block

CTLMode WP_INTEST_MODE {
  TestMode InternalTest;
      DomainReferences {
          MacroDefs ex_wrapper_macros;
          ScanStructures EX_wrapper_WSP_chains
                           EX_wrapper_WPP_chains;
      } // DomainReferences block
  Internal {
      all_functional {DataType Functional;}
      wpp_mbist {DataType Unused;}
  } // Internal block
  PatternInformation {
      Pattern WpIntestSetupPattern {
      Purpose EstablishMode;
      Protocol Macro instruction_mode_setup;
      } // Pattern block

      Pattern WpIntestRunPattern {
          Purpose Scan;
          Protocol Macro run_wp_intest;
      } // Pattern block
```

```
            Macro run_wp_intest {
                Purpose DoTest;
                UseByPattern Scan;
            } // Macro block

            Macro instruction_mode_setup {
                Purpose ModeControl;
                UseByPattern EstablishMode;
            } // Macro block
        } // PatternInformation block
        TestModeForWrapper WP_INTEST 0100;
    } // CTLMode WP_INTEST_MODE block

    CTLMode WS_INTEST_MODE {
        TestMode InternalTest;
        DomainReferences {
            MacroDefs ex_wrapper_macros;
            ScanStructures EX_wrapper_WSP_chains;
        } // DomainReferences block
        Internal {
            all_functional {DataType Functional;}
            wpp_mbist {DataType Unused;}
        } // Internal block
        PatternInformation {
            Pattern WsIntestSetupPattern {
                Purpose EstablishMode;
                Protocol Macro instruction_mode_setup wrapper_reset;
            } // Pattern block
            Pattern WsIntestRunPattern {
                Purpose Scan;
                Protocol Macro run_ws_intest;
            } // Pattern block

            Macro instruction_mode_setup {
                Purpose ModeControl;
                UseByPattern EstablishMode;
            } // Macro block

            Macro run_ws_intest {
                Purpose DoTest;
                UseByPattern Scan;
            } // Macro block

        } // PatternInformation block
        TestModeForWrapper WS_INTEST 0010;
    } // CTLMode WS_INTEST_MODE block
```

```
CTLMode WP_INTEST_MBIST_MODE {
  TestMode InternalTest;
  DomainReferences {
     MacroDefs ex_wrapper_macros;
     ScanStructures EX_wrapper_WSP_chains
                    EX_wrapper_WPP_chains;
  } // DomainReferences block
  Internal {
     all_functional {DataType Functional;}
     wpp_scan {DataType Unused;}
  } // Internal block
  PatternInformation {
     Pattern WpIntestMbistSetupPattern {
        Purpose EstablishMode;
        Protocol Macro instruction_mode_setup;
     } // Pattern block
     Macro instruction_mode_setup {
        Purpose ModeControl;
        UseByPattern EstablishMode;
     } // Macro block
  } // PatternInformation block
TestModeForWrapper WP_INTEST_MBIST 0111;
} // CTLMode WP_INTEST_MBIST_MODE block

CTLMode WP_EXTEST_MODE {
  TestMode Bypass Normal;
  DomainReferences {
     MacroDefs ex_wrapper_macros;
     ScanStructures EX_wrapper_WSP_chains
                    EX_wrapper_WPP_chains;
  } // DomainReferences block
  Internal {
     all_functional {DataType Functional;}
     wpp_mbist {DataType Unused;}
  } // Internal block
  PatternInformation {
     Pattern WpExtestSetupPattern {
        Purpose EstablishMode;
        Protocol Macro instruction_mode_setup;
     } // Pattern block

     Macro instruction_mode_setup {
        Purpose ModeControl;
        UseByPattern EstablishMode;
     } // Macro block
  } // PatternInformation block
  TestModeForWrapper WP_EXTEST 0110;
} // CTLMode WP_EXTEST_MODE block
```

```
CTLMode WS_SAFE_SINGLE_MODE {
  TestMode Isolate;
  DomainReferences {
      MacroDefs ex_wrapper_macros;
      ScanStructures EX_wrapper_WSP_chains;
  } // DomainReferences block
  Internal {
      all_functional {DataType Functional;}
      wpp_mbist {DataType Unused;}
} // Internal block
  PatternInformation {
      Pattern WsSafeSingleSetupPattern {
          Purpose EstablishMode;
          Protocol Macro instruction_mode_setup;
      } // Pattern block
      Macro instruction_mode_setup {
          Purpose ModeControl;
          UseByPattern EstablishMode;
      } // Macro block
  } // PatternInformation block
  TestModeForWrapper WS_SAFE_SINGLE 0011;
  } // CTLMode WS_SAFE_SINGLE_MODE block

CTLMode WP_EXTEST_SEQ_MODE {
  TestMode ExternalTest;
  DomainReferences {
      MacroDefs ex_wrapper_macros;
      ScanStructures EX_wrapper_WSP_chains
                         EX_wrapper_WPP_chains;
  } // DomainReferences block
  Internal {
      all_functional {DataType Functional;}
      wpp_mbist {DataType Unused;}
  } // Internal block
  PatternInformation {
      Pattern WpExtestSeqSetupPattern {
          Purpose EstablishMode;
          Protocol Macro instruction_mode_setup;
      } // Pattern block
      Macro instruction_mode_setup {
          Purpose ModeControl;
          UseByPattern EstablishMode;
      } // Macro block
  } // PatternInformation block
  TestModeForWrapper WP_EXTEST_SEQ 1001;
  } // CTLMode WP_EXTEST_SEQ_MODE block
```

```
<10.853>  CTLMode WP_INTEST_SEQ_MODE {
<10.854>    TestMode InternalTest;
<10.855>    DomainReferences {
<10.856>        MacroDefs ex_wrapper_macros;
<10.857>        ScanStructures EX_wrapper_WSP_chains
                               EX_wrapper_WPP_chains;
<10.858>    } // DomainReferences block
<10.859>    Internal {
<10.860>        all_functional {DataType Functional;}
<10.861>        wpp_mbist {DataType Unused;}
<10.862>    } // Internal block
<10.863>    PatternInformation {
<10.864>        Pattern WpIntestSeqSetupPattern {
<10.865>            Purpose EstablishMode;
<10.866>            Protocol Macro instruction_mode_setup;
<10.867>        } // Pattern block
<10.868>        Macro instruction_mode_setup {
<10.869>            Purpose ModeControl;
<10.870>            UseByPattern EstablishMode;
<10.871>        } // Macro block
<10.872>    } // PatternInformation block
<10.873>    TestModeForWrapper WP_INTEST_SEQ 1010;
<10.874>    } // CTLMode WP_INTEST_SEQ_MODE block
<10.875>
<10.876>  } // Environment EX_wrapper block
<10.877>
<10.878>  Pattern WsBypassSetupPattern {
<10.879>    P {"WSI"=0000;}
<10.880>  }
<10.881>  Pattern WpBypassSetupPattern {
<10.882>    P {"WSI"=0101;}
<10.883>  }
<10.884>  Pattern WsExtestSetupPattern {
<10.885>    P {"WSI"=0;}
<10.886>    P {"WSI"=0;}
<10.887>    P {"WSI"=0;}
<10.888>    P {"WSI"=1;}
<10.889>  }
<10.890>  Pattern WpIntestRunPattern {
<10.891>    P {"WPSI[0..3]"=0000; "WPSO[0..3]"=XXXX;}
<10.892>    P {"WPSI[0..3]"=0100; "WPSO[0..3]"=XXXX;}
<10.893>    P {"WPSI[0..3]"=0011; "WPSO[0..3]"=XXXX;}
<10.894>    P {"WPSI[0..3]"=0000; "WPSO[0..3]"=XXXX;}
<10.895>    P {"WPSI[0..3]"=0110; "WPSO[0..3]"=XXXX;}
<10.896>    P {"WPSI[0..3]"=0000; "WPSO[0..3]"=XXXX;}
<10.897>    P {"WPSI[0..3]"=0011; "WPSO[0..3]"=XXXX;}
<10.898>  } // WpIntestRunPattern block
```

```
Pattern WsIntestRunPattern {
  P {"WSI"=1; "WSO"=X;}
  P {"WSI"=0; "WSO"=X;}
  P {"WSI"=0; "WSO"=X;}
  P {"WSI"=1; "WSO"=X;}
  P {"WSI"=1; "WSO"=X;}
  P {"WSI"=0; "WSO"=X;}
  P {"WSI"=1; "WSO"=X;}
} // WsIntestRunPattern block

Pattern WpIntestSetupPattern {
  P {"WSI"=0100;}
}
Pattern WsIntestSetupPattern {
Setup { }
  P {"WSI"=0010;}
}
Pattern WpIntestMBistSetupPattern {
  P {"WSI"=0111;}
}
Pattern WpExtestSetupPattern {
  P {"WSI"=0110;}
}
Pattern WsSafeSingleSetupPattern {
  P {"WSI"=0011;}
}
Pattern WpIntestSeqSetupPattern {
  P {"WSI"=1010;}
}
Pattern WpExtestSeqSetupPattern {
  P {"WSI"=1001;}
}
```

Chapter 11
SOC Integration of 1500 Compliant Cores

11.1 SOC integration

The 1500 standard targets ease of integration of cores into system chips while maintaining good testability. In general, the number of IC pins available to accommodate test signals is known. Within that constraint, a trade-off between test time and silicon area can be made while defining the SOC level test architecture. For example, if pre-computed tests of a certain core only need to transport small numbers of bits, the core only needs a narrow Test Access Mechanism (TAM) to transport this test data. In this case it is often better to devote the scarce IC pins and silicon area to other TAMs that connect to cores that require more bandwidth during test.

This chapter addresses core integration considerations and shows different architectures that connect multiple compliant wrapped cores inside an SOC. Most of the discussion is limited to wrapped cores since unwrapped cores will be wrapped as part of the integration process and as such do not form a special case that requires discussion in this chapter. The handling of functional connections between cores is not in the scope of this book and will not be addressed here.

11.1.1 1500 compliant wrapped core types.

Figure 113 (page 250) shows three types of wrapped cores. The first core (A) is considered delivered from a third party (3p) and consists of core logic surrounded by a 1500 compliant wrapper. This type of core can be a hard core that is fully prepared for test and is delivered with the test set required for manufacturing included. In order to develop an efficient and cost effective test set for the complete SOC it is helpful to be able to request this type of core with an on-core test infrastructure that fits the SOC test architecture. Embedding generic off-the-shelf cores often results in test time inefficiency as will be explained later.

The second type of wrapped core (B) shown in Figure 113 is built with a third party unwrapped core and a wrapper that is created during the core integration process. In comparison to Core (A) discussed earlier, this type of core can be 'tuned' to the SOC environment more easily. If necessary, the SOC integrator can adapt the core test interface to the SOC test access mechanism. This may be needed if test of the embedded core can only use a narrow test access mechanism because of the limited availability of SOC pins. Re-use of pre-defined test sets for this type of core is possible, but needs translation from the unwrapped core boundary to the SOC boundary. This can be a tedious task if time multiplexing is used to reduce the amount of SOC pins involved during test.

The third type of core (C) shown in Figure 113 is not third party logic and could result from the use of 1500 to implement a 'divide-and-conquer' test strategy for the entire SOC. The core and wrapper test infrastructure, in this case, can be tuned completely to the SOC test architecture. This flexibility presents the best opportunity to create an efficient SOC test set.

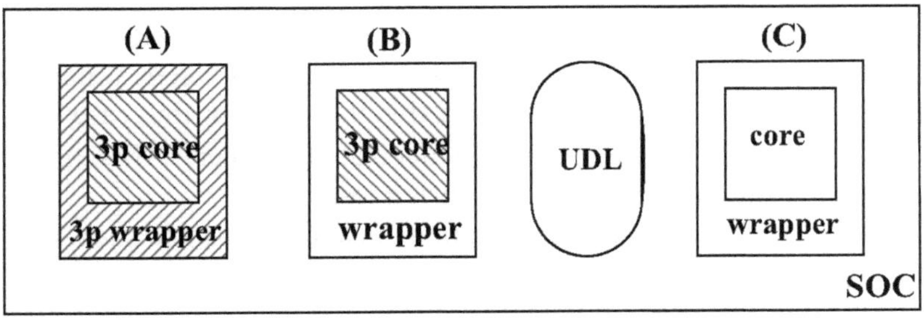

Figure 113 SOC with three types of 1500 compliant wrapped cores

Three types of 1500 compliant cores can be identified based on their types of test interfaces. These various types are depicted in Figure 114 as core (A), (B) and (C) and are further explained in the following paragraphs.

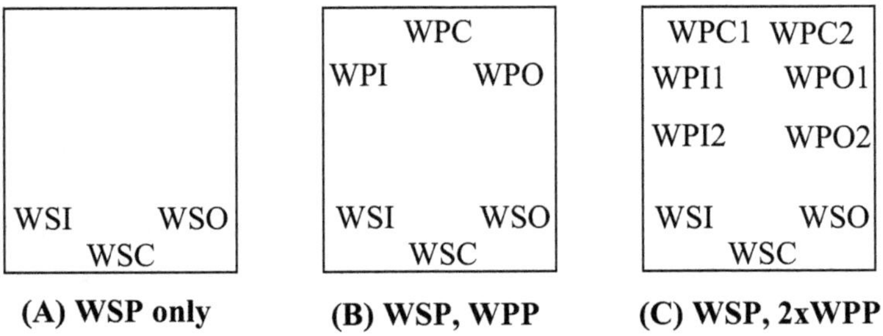

Figure 114 Example 1500 compliant wrapper interfaces

Wrapped core (A) implements only a serial test interface. This means that the test data volume that is delivered for internal test of the core must be small. Under the assumption that high quality tests are delivered this can be a result of either very little logic that needs to be tested, which is often the case for analog/mixed signal IP, or self-test has been implemented to reduce the data volume that must be communicated to and from the SOC boundary. For this type of wrapped core a single wire can be used to access the WBR cells involved in interconnect testing. Access to these types of cores needs to be distributed over the available IC pins to have enough bandwidth available in case a large pattern set is needed to test the UDL between wrapped cores. An alternative approach is to limit the number of interconnect patterns by adding a wrapper to the UDL. This leaves only wire interconnect, at the SOC level, which can be tested efficiently with only a few patterns.

Wrapped core (B), shown in Figure 114 (page 250), has both the mandatory serial WSP and a parallel WPP. For digital cores this typically means that the core is tested with either scan tests with high data volume or functional tests that require terminals to be directly observed or controlled from SOC pins.

Wrapped core (C), see Figure 114, contains two parallel ports (WPPs) in its terminal interface. There is no restriction on the number of WPPs used by a core.

Beyond the different types of wrapped cores, the SOC designer determines the best way to connect multiple WSPs and WPPs to the SOC's TAMs. This decision can be tailored to specific SOC test requirements as explained in the following sections.

11.1.2 Core connection types

On-chip access to wrapped cores embedded in an SOC is arranged by TAMs. A TAM is used to transport stimuli from a test pattern source to a core-under-test, and to transport test responses from the core-under-test to a test pattern sink. Source and sink can be on-chip structures such as Built-In Self Test (BIST) engines or off-chip test engines such as an Automated Test Equipment (ATE). This section introduces three basic core interfaces to TAM connection types. Although one core often uses the same type of TAM for stimulus as well as response transportation, this is not required and various combinations may co-exist.

Figure 115 shows the following three types of core interface connections:

- Distribution (A)
  Each core is assigned to a number of SOC pins such that the total number of SOC pins available for testing is divided between all of the

cores. In general, cores requiring huge test bandwidth utilize most of the SOC pins assignments. Cores that transport only a small amount of test data, for example BISTed cores, require only a few pins.

- Daisy-chaining (B)
  A fixed width TAM is concatenated through all cores, with a bypass per core sharing the same width. This type of connection best fits interfaces that have an equal amount of test data inputs and outputs such as a scan test interface. The maximum width can equal half the total number of SOC test data pins available.
- Multiplexing (C)
  Each core connects to the same set of SOC pins via a dedicated TAM. This pin resource sharing is controlled by multiplexers which are controlled from SOC level test mode bits. Only one test mode is active at a time and each test mode enables the connection of one core to the SOC pins.

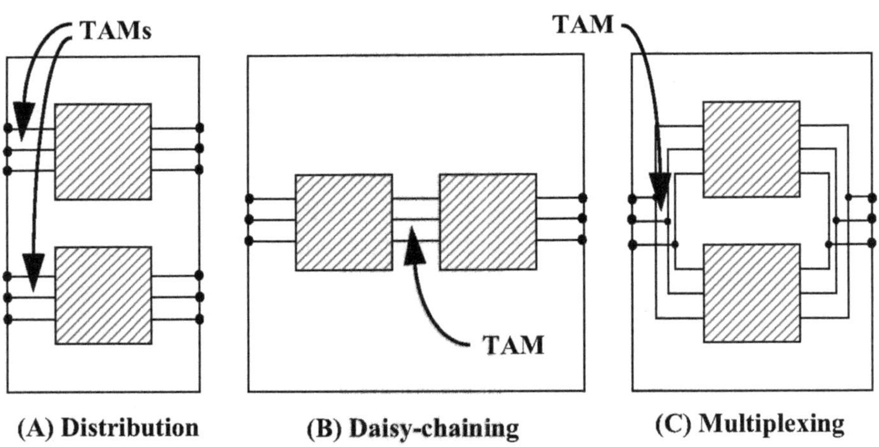

Figure 115 Interface connection types

In the distribution architecture (A) the total available SOC TAM width is distributed over the wrapped modules. The optimal number of wires that is assigned to a module depends on the test data volume and scan chain design of all embedded modules in the SOC. As a consequence, this architecture is sensitive to even small changes in the SOC design. These can result in huge test time inefficiencies. In addition, to minimize the overall test time for multi-core scan testing, each core needs its own scan control. Sharing scan control between cores can again lead to an increase in test time for some of the cores involved. In the daisy-chaining architecture (B) all modules get access to the full available SOC TAM width. Due to the bypass mechanism for each embedded core, any subset of wrappers can be accessed simultaneously. As a result cores can be tested individually in a sequential test schedule or in parallel with other cores. Simultaneous access of multiple cores also enables interconnect test between cores. In the multiplexing architecture (C) interconnect testing between cores is cumbersome because only one core is accessed at a time.

11.1.3 Connecting 1500 compliant cores

The 1500 standard does not prescribe a specific type of interface connection at the SOC level. The connection types listed in Section *11.1.2* (see page 251) can be used for both WSP and WPP interfaces. A single SOC might even contain one or more TAMs, each using a combination of distribution, daisy-chaining and/or multiplexing architecture. Designing the SOC TAM architecture is often presented as finding the optimal number of TAMs, their widths and the cores assigned to them with respect to various cost factors such as test time, area overhead, power dissipation, etc. In general, a wider TAM provides more bandwidth at the cost of more wiring area. When an IC has only a few pins and the test patterns are to be transported to and from an external ATE, a wide access mechanism does not make much sense. The TAM bandwidths in combination with the test data volumes of the individual cores and their test schedule result in the final test time and power dissipation figures. This section gives insight into the design of TAMs for connecting multiple 1500 compliant cores.

Integrating a 1500 compliant wrapped core in a larger system chip always requires the mandated WSP. In Figure 116 (page 254) two examples of WSP interface connections are shown. The cores do not have a WPP which means that the maximum bandwidth per core consists of a single wire for test data. Distributing the WSPs as shown in Figure 116(A) results in the maximum achievable bandwidth. Tests that are delivered with the cores can be translated to SOC chip pins easily with this WSP distribution. The connections between SOC pins and wrapper terminals consist of wiring only. This means that the translation of test protocols and accompanying instruction loading from the wrapper boundary to SOC pins can easily be done by mapping terminal names to chip pin names. Daisy-chaining the WSPs, as shown in Figure 116(B), results in a single serial interface port at the SOC level that complies to a 1500 WSP. In this scenario, translation of core internal tests cannot be done simply by mapping terminal names to chip pins. Not all WSP terminals are connected directly to chip pins. The WSO of the left core (C1) feeds through a register of the right core (C2) towards the SOC output pin. Similarly, data meant for the WSI of the right core (C2) has to traverse a register of the left core (C1). This means that both the protocol for instruction loading and the protocol for test pattern execution of each core need to be transformed such that the proper sequential depth is supported in their serial test data access path.

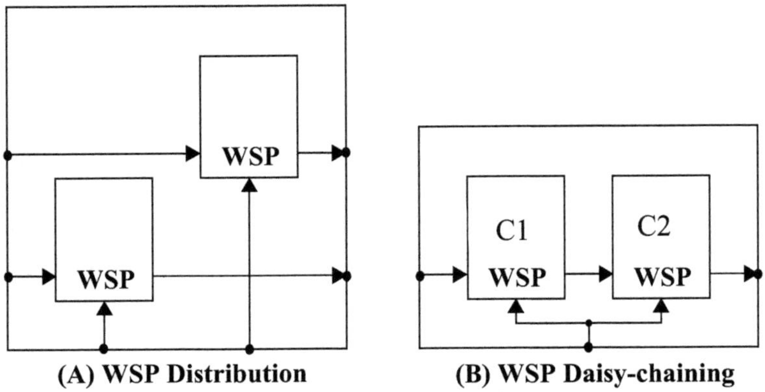

Figure 116 WSP distribution and daisy-chaining

The capability of the 1500 compliant wrapped cores depicted in Figure 116 was extended with a multi-bit WPP to add more bandwidth for both core internal and core interconnect test. Figure 117 shows the resulting core connections when the WPPs of both cores are distributed over SOC pins. Here again the reuse of core internal tests is most simple whenever there is a full distribution of both the WSP and WPP, as shown in Figure 117(A). Translation of both testmode initialization sequences and test protocols can be done by mapping terminal names to SOC pins. Apart from the simple translation benefit, direct access to wrapper terminals also enables the reuse of nonscan tests (e.g. MBIST). Distributing WPPs while daisy-chaining WSPs as shown in Figure 117 (B) results in simple test protocol translation as described above for parallel instructions only. Reusing nonscan tests in this architecture is possible only if the testmode initialization sequence is kept separate from the nonscan test protocol. A daisy-chain architecture that uses a sequential bypass mechanism instead of direct access, by definition, requires a more complicated protocol transformation. Compiling the protocol for instruction loading and creating the proper initialization sequences for test mode setup is relatively easy for daisy-chained WSPs.

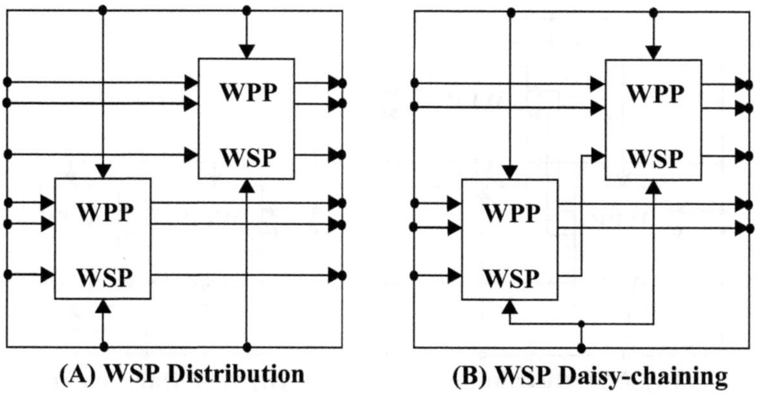

Figure 117 WPP distribution

Instead of distributing WPPs across the SOC pins, the WPPs can be connected to a single TAM by daisy-chaining them as shown in Figure 118. This type of connection is typically used for scan test interfaces. Other types of tests, such as functional tests, generally require direct access that is not provided by the sequential bypass paths used in a daisy-chaining architecture. Implementing these bypass paths in this type of architecture without using storage elements often results in too much delay. As explained in Section *11.1.2* (see page 251) the benefit of daisy-chaining architecture over the distribution scheme is that the implementation of the daisy-chain architecture can be done without any knowledge of the test data volume information of any of the cores. The maximum bandwidth is always utilized in this scenario. However, distributing TAM wires over modules is a challenge and can only be done with precision once all test data for all wrapped modules is available. In practice this means that test architecture design that makes use of the distribution scheme cannot be completed until very late in the design cycle, often close to completion.

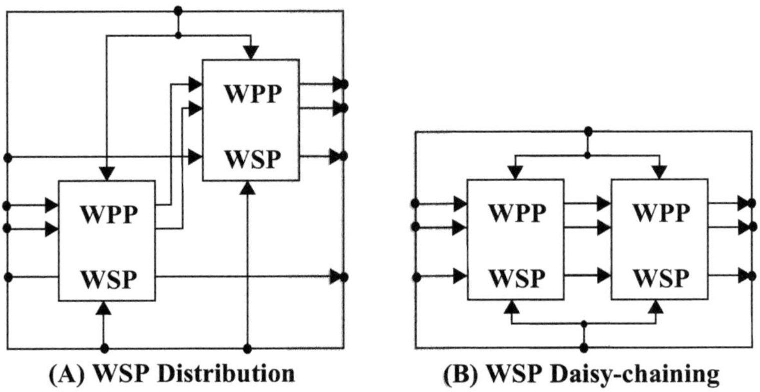

Figure 118 WPP daisy-chaining

Wrapped cores that are not fully BISTed may come with more than one WPPs which can be used for scan test, functional test and any specific tests for non-standard logic modules that have been embedded into the wrapped core. This type of core often has special requirements during test. For example, direct access can be required, but also environmental requirements such as power isolation might result in test protocols that cannot run in parallel with other tests. For this type of WPP, the use of multiplexing, as shown in Figure 119, is most natural. In Figure 119, the core on the left side(C1) has a WPP1 which requires direct access. The benefit of multiplexing is that the full SOC bandwidth can be used and all types of tests, including functional tests, can be reused without any complicated translation. Testing the interconnect between wrapped cores is done without using the multiplexing access. In fact the multiplexers and accompanying paths are part of the UDL and tested together with other logic during core external testing.

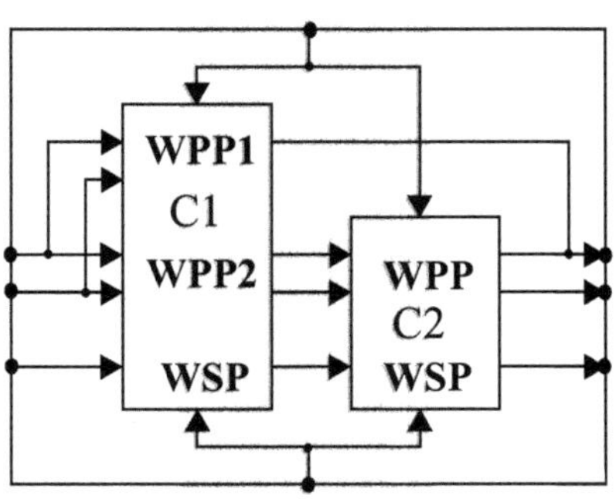

Figure 119 WPP Multiplexing

None of the scenarios discussed so far address the case where the number of chip pins that can connect to a TAM is smaller than the number of terminals of a WPP that need connection to the same TAM. In this case, the WPP width needs to be adapted to the TAM width, if possible. For instance, a scan test interface can easily be adapted because of the time multiplexing nature of the protocol. When a WPP has direct control and observe requirements, adaptation might not be possible. Adapting a WPP to a TAM width typically occurs when scan inserted third party cores are embedded in low pin count SOCs. In Figure 120, WPP width adaptation is shown for two types of third party cores. In the case of wrapped core (A) this type of width change can be done only as part of the core integration process. In the case of an unwrapped core, logic can be added in the wrapper to implement the same capability. Wrappers can also support multiple TAM widths via programmable WPP configuration logic inside the wrapper.

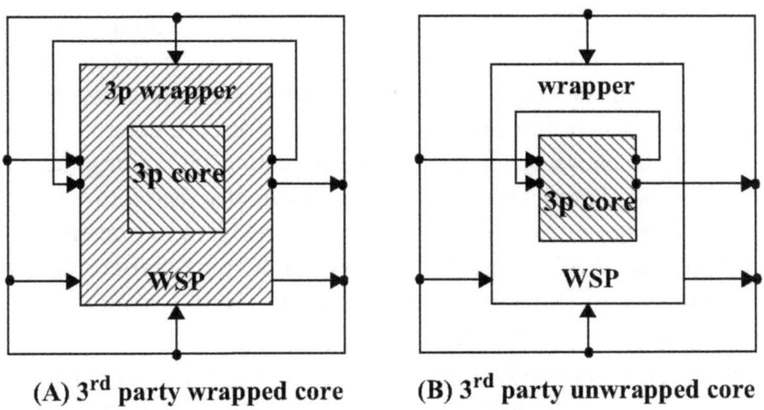

Figure 120 WPP width adaptation

11.1.4 Interfacing to SOC pins

The connection of cores to TAMs and the SOC boundary was discussed in Section *11.1.2* (see page 251). However, the fact that SOC pins are scarce resources, that cannot simply be claimed for test usage only, has not been addressed yet. Figure 121 shows that functional SOC pins can be reused for test access to WSP terminals. A single dedicated test pin *RST* is used to enable functional operation or test operation in the SOC. While active, this pin forces all WIRs into reset state and forces the pin reduction logic to connect to functional output paths instead of wrapper WSO terminals.

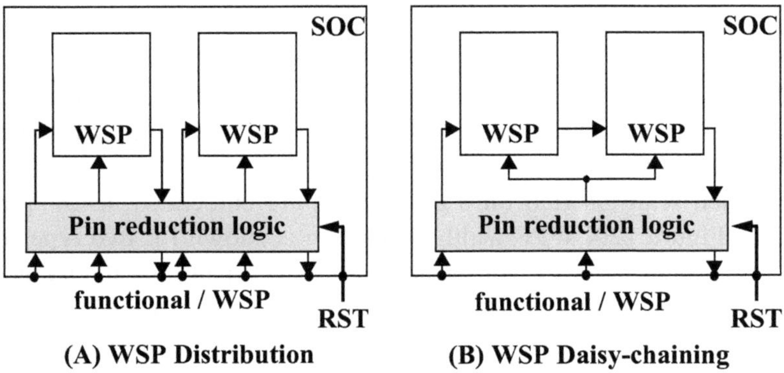

Figure 121 Reusing functional pins for WSP access

Care should be taken with pin reduction schemes involving shared functional pins for WSP terminals. Each test places certain restrictions on the test control pins. For instance, the SelectWIR signal requires a continuous low value during serial scan test protocol execution. For core external testing this fixed value might cause coverage loss while testing UDL. In general, functional pins that are shared with WSP pins must directly feed into wrapper cells to limit the coverage impact. A more robust approach is to prevent the use of functional pins for WSP access and use only SOC pins that are dedicated to test. For large system chips these pins are easily found when the SOC interface contains an 1149.1 JTAG TAP port.

Instead of pin reduction logic that interfaces WSP terminals to SOC pins a TAP controller can be used. Figure 122 shows example connections of WSP terminals to a TAP controller. In both examples, the WSP registers operate as JTAG data registers. The WSPs can be connected in a distributed (A) form which means that each wrapper acts as a JTAG register on its own and is addressed by its own set of JTAG instruction opcodes. WSPs can also be daisy-chained (B). This forms a single JTAG register that is addressed by a single set of JTAG opcodes. The wrappers shown in Figure 122 have only a WSP. This means that both WIR loading and test protocol execution are controlled by the TAP controller and its standard Finite State Machine (FSM) protocol. Transformation of initialization and test protocol from the wrapper boundary to the SOC pins via the TAP controller is guaranteed by the plug-and-play properties of the 1500 standard. Serial instructions require hold behavior while no WSP event is active. This hold behavior is needed to keep test data valid while traversing FSM states that do not activate an event in a standard JTAG protocol.

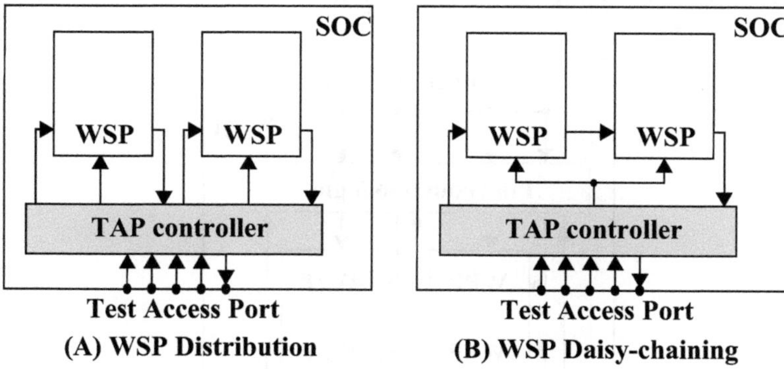

Figure 122 Single TAP controller WSP access

Readers familiar with 1149.1 might recall that JTAG data registers must be of fixed length when addressed by instructions that are not flagged private. The WSP access in Figure 122 fulfills this requirement only when SelectWIR is active in JTAG shift data register state. This selects the fixed length WIR register for 1500 instruction loading. All other 1500 register lengths depend on instructions loaded into the WIR instead of the JTAG instruction register. This means they can only be addressed by private JTAG instructions in the architecture shown in Figure 122.

One can conclude from the above that accessing WSPs being addressed as JTAG data registers makes the most sense if the WSP is used for WIR access only. This means that in order to be 1149.1 compliant, the SOC should be tested without the use of any 1500 serial instructions. For the SOC this means that it should be completely tested without the use of any serial 1500 instruction. Note that WS_EXTEST will be available on every core. If there is a need to exercise this instruction, a private instruction would be required in order to control the WDRs with the WSP from the TAP controller. Figure 123 and Figure 124 show example SOC test architectures that fit this approach. Both show an SOC that has two wrapped cores embedded, each having a serial and parallel test port. For clarity, these architectures are constructed under the assumption that only parallel instructions are used to test the complete SOC. The TAP controller in this type of architecture is used solely for loading instructions into the wrappers. The execution of test protocols is entirely done via the WPPs that are connected to functional SOC terminals via TAMs and pin reduction logic. The only difference between the examples shown is a different control mechanism for the pin reduction logic.

In Figure 123 the pin reduction logic configuration is directly controlled from the TAP controller, so directly decoded from the JTAG instruction register. This means that, for this architecture, switching test modes involves loading different JTAG instructions and loading different 1500 instructions.

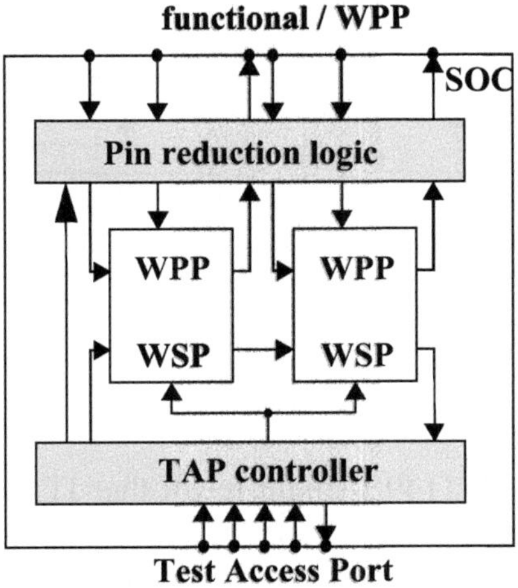

Figure 123 Reusing functional pins for WPP access

Figure 124 shows an alternative control mechanism for the pin reduction logic. A WIR is added to take care of test control at SOC level. The benefit of this approach is that after loading the appropriate JTAG instruction switching test modes involves loading WIR instructions only. Control at the SOC and core level share the same mechanism to configure the environment and access for test.

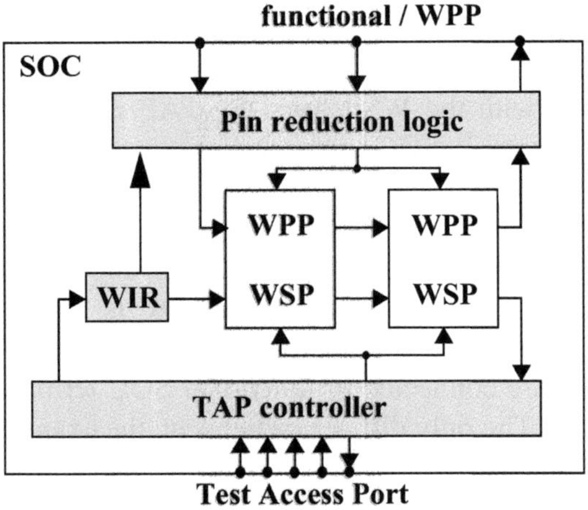

Figure 124 Using the WIR for SOC test control

Another approach for interfacing 1500 wrappers to an SOC TAP is shown in Figure 125. The architecture presented in this figure has multiple TAP controllers. On the left hand side, the regular controller that implements the standard capabilities for board testing is shown, as well as the SOC test control. The other two TAP controllers are created as part of the core integration process. The addition of a TAP state machine to control the WSP of a 1500 wrapper transforms this wrapper into a JTAG architecture. The WIR serves as its instruction register and the serial data registers serve as JTAG data registers. Similar to daisy-chaining 1500 WSPs, concatenating multiple TAPs while sharing their control creates a single compliant TAP.

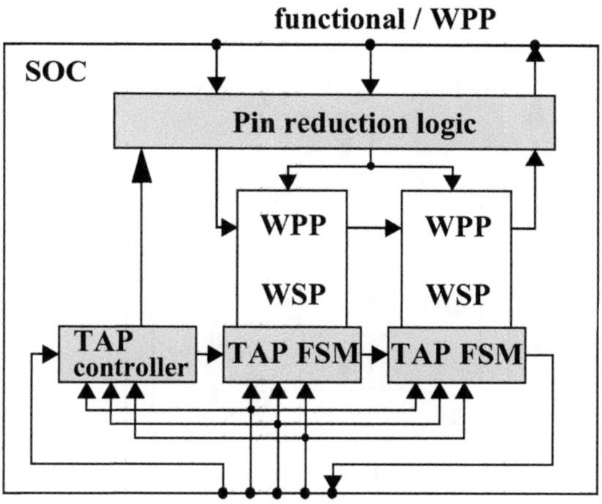

Figure 125 Multiple TAP controller architecture

The Multiple TAP controller architecture fully complies to 1149.1 and does not need private JTAG instructions when wrapper serial data registers need to be addressed. This is a result of operating the WIR as part of the SOC JTAG instruction register. A drawback of the architecture shown in Figure 125 is the length of the JTAG instruction register that grows with the amount of concatenated WIR bits. Minimizing JTAG instruction register lengths can be addressed by hierarchical TAP architectures that can be found in other literature.

11.2 TAP controller interface

The 1500 standard does not prescribe a standard interface between the WSP and a JTAG TAP controller but this standard was designed to allow interface compatibility with a TAP controller as introduced in Section *11.1.4* (see page 257).

11.2.1 Interfacing TAP FSM to WSC

All the events supported by the WSC interface of a 1500 wrapper correspond directly to the events defined in the 1149.1 TAP FSM shown in Figure 126, with the exception of Transfer event. The Transfer event is discussed separately in Section *11.2.3* (see page 266).

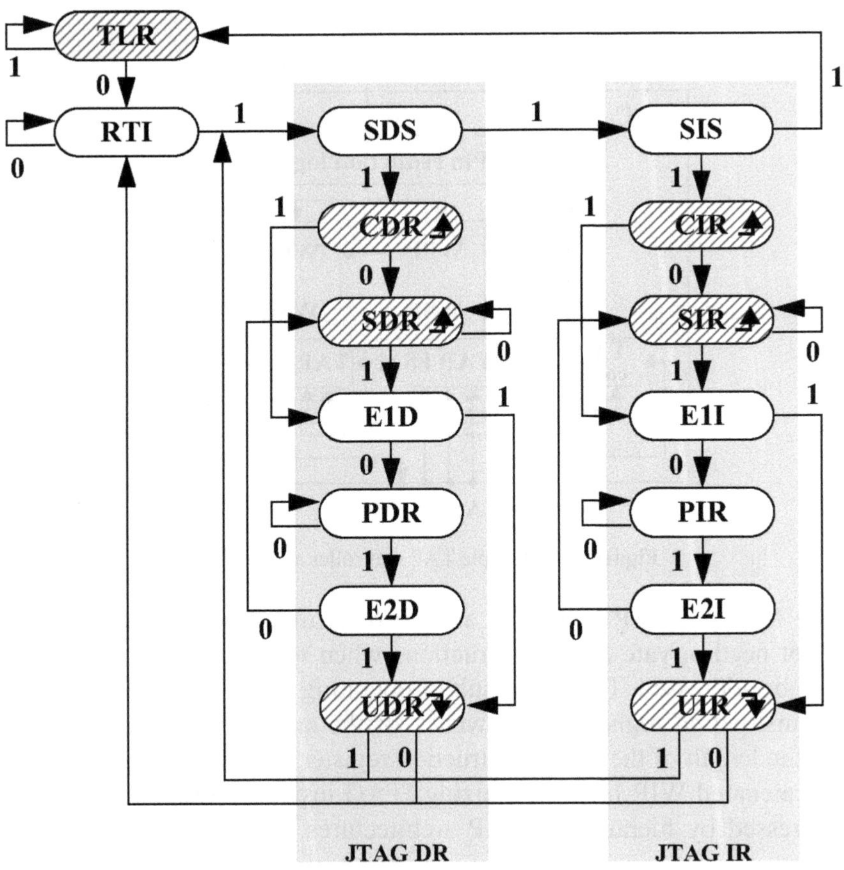

Figure 126 IEEE Std. 1149.1 TAP FSM states

The TAP FSM contains 16 states that can be traversed using a dedicated test clock TAP TCK and a test mode select signal TAP TMS. The value shown adjacent to each state transition in Figure 126 represents the signal value at TMS at the rising edge of TCK. Two shaded boxes identify TAP states that control either Data Register (JTAG DR) or Instruction Register (JTAG IR) scan operations. The hashed states can be used to trigger WSC events such that the 1500 registers comply to the behavior defined in 1149.1. Details are shown in Figure 127. The Test Logic Reset (TLR) state forces a design into mission mode. Accordingly, 1500 wrappers must be forced into disabled state. The Run Test Idle (RTI) state is not involved in register access, but is used as an idle state during test. When interfacing to a 1500 wrapper the RTI state is typically active while a WPP is used during test.

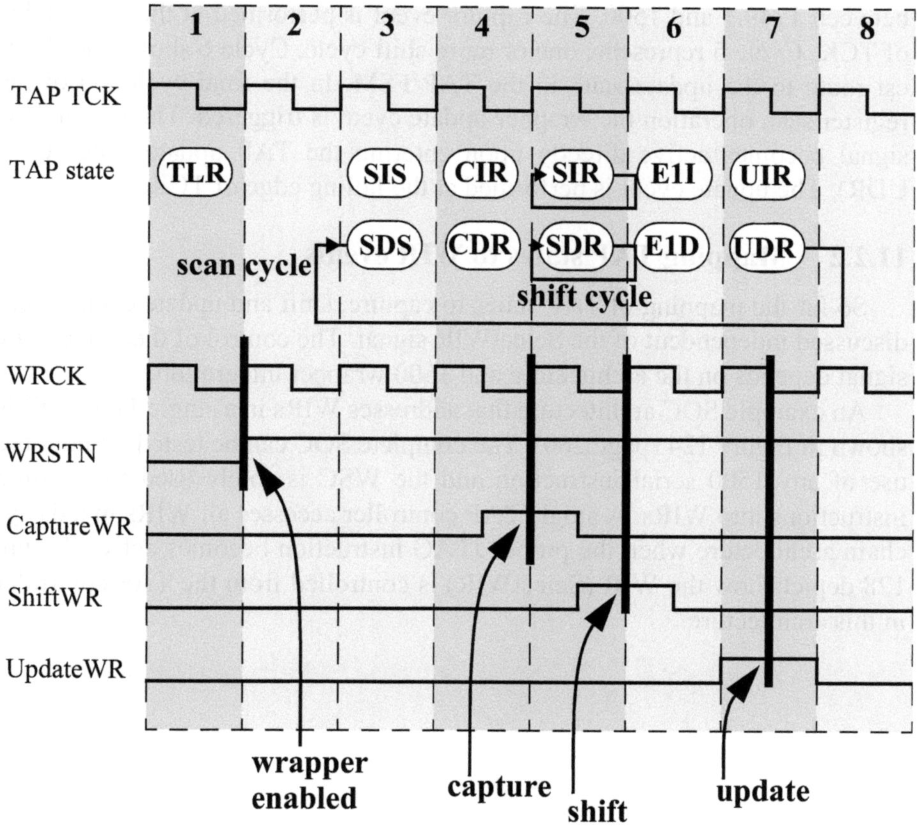

Figure 127 Mapping TAP states to WSC control

In Section *11.1.4* (see page 257) two flavors of interface between TAP and WSP were discussed. The first implementation controls the WSC during JTAG DR operations only. The second selects the WIR during JTAG IR operations and WDRs during JTAG DR operations. Figure 127 shows the mapping between TAP states and WSC control for each case. The SelectWIR control is different for both cases and is shown in Section *11.2.2* (see page 264). In cycle 1 the TLR state drives the WRSTN to its active value which forces the 1500 wrapper into its disabled state. The wrapper is enabled by leaving the TLR state. The wrapper clock WRCK can be directly derived from the TAP TCK. Cycle 2 represents multiple states that do not trigger WSC events. Cycles 3 to 7 show a typical register scan operation for both JTAG IR and JTAG DR types of register. Cycle 3 represents a register select cycle that does not trigger any WSC event. The capture state for the selected register (JTAG IR or JTAG DR) is active in cycle 4. The CaptureWR signal changes at the falling edge of TCK to allow for a robust timing interface between 1149.1 and 1500. The capture event is performed at the rising edge of TCK. Cycle 5 represents one or more shift cycle. Cycle 6 shows the shortest route to the update state in the TAP FSM. In the final cycle (7) of the register scan operation the wrapper update event is triggered. The UpdateWR signal becomes active directly upon entering the TAP update state (UIR/UDR). The update event is performed at the falling edge of TCK.

11.2.2 Mapping TAP states to WIR events

So far the mapping of TAP states to capture, shift and update events were discussed independent of the SelectWIR signal. The control of the SelectWIR signal depends on the architecture and 1500 wrapper integration.

An example SOC architecture that addresses WIRs in a single JTAG DR is shown in Figure 124 (page 260). The complete SOC can be tested without the use of any 1500 serial instruction and the WSC is solely used for loading instructions into WIRs. A single TAP controller accesses all WIRs in a daisy-chain architecture when the proper JTAG instruction becomes active. Figure 128 depicts how the WIR (SelectWIR) is controlled from the TAP controller in this architecture.

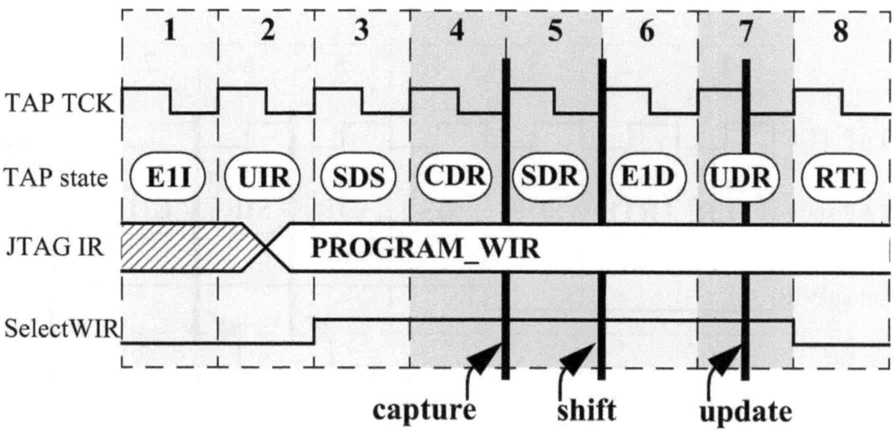

Figure 128 WIR events when controlled as JTAG data register

In cycle 1 and 2 the loading of the PROGRAM_WIR instruction opcode into the JTAG IR is finalized. This JTAG instruction is used to load 1500 instruction opcodes into the daisy-chained WIRs. During other JTAG instructions the SelectWIR signal is continuously low. The PROGRAM_WIR instruction becomes active on the falling edge of TCK during the Update Instruction Register (UIR) state in cycle 2. The WIR chain is now selected as the JTAG DR and the accompanying operations are passed to the WIR by activating the SelectWIR signal. The SelectWIR is made active during all JTAG DR states shown in Figure 126 (page 262). The JTAG DR states shown in cycles 3 to 7 of Figure 128 (page 265) represent a typical WIR scan operation. In cycle 8, the WIRs have been loaded and tests can be executed. During these tests the TAP FSM is parked in the RTI state which can drive SelectWIR to the inactive state.

An example SOC architecture that uses a daisy-chain of TAP controllers for test access is shown in Figure 125 (page 261). In this architecture all 1500 wrapper WSPs are transformed into 1149.1 compliant TAPs by adding a TAP state machine. The WIRs are operated as JTAG IR and controlled accordingly by the JTAG IR operations of the TAP FSM shown in Figure 126 (page 262). The JTAG DR operations are used to control selected WDRs. Figure 129 depicts how the WIR (SelectWIR) is controlled from the TAP FSM. The SelectWIR signal is active during all JTAG IR states in order to pass the capture, shift and update events to the WIR.

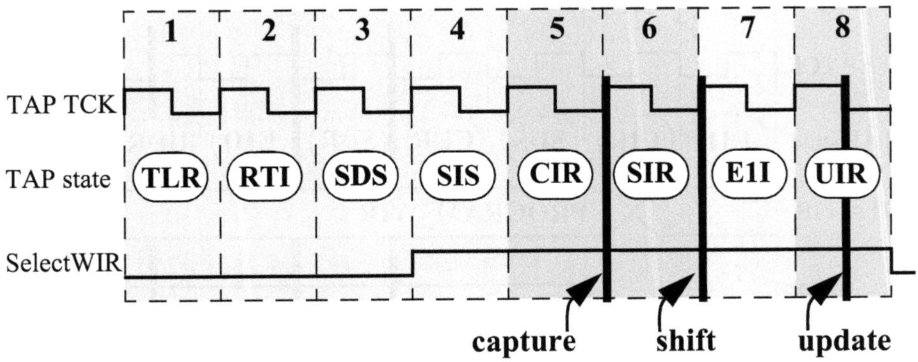

Figure 129 WIR events when controlled as JTAG Instruction Register

11.2.3 Mapping TAP states to WBR cell events

The WSP is used for WIR access and also gives access to WDRs. The events that control the data register operations can be typically mapped to JTAG DR states in the presence of a TAP controller interface. Example WDRs are the WBY and WBR. Figure 130 shows a typical scan cycle and its WBR cell events. The WBR cells hold their value in the absence of an event (cycles 1 and 4). The SelectWIR signal is driven low during all JTAG DR states to guarantee WDR access.

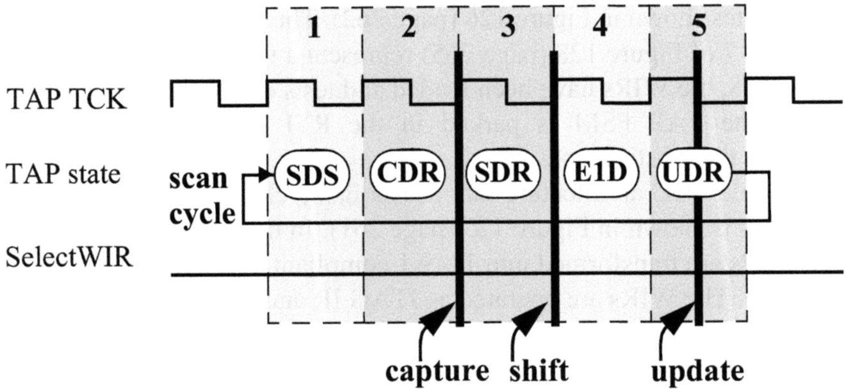

Figure 130 WBR cell events when controlled as JTAG data register

11.2.3.1 Mapping TAP states to the WBR cell Transfer event

So far, the optional Transfer event has not been discussed with respect to a WSP that is controlled from a TAP controller.

In Section *6.3.1.4* (see page 139) two possible data movements for the Transfer event are identified. One is transferring data from an off-shift-path storage element to an on-shift-path storage element. The second is the transfer of data from an on-shift-path element to another on-shift-path element within a single WBR cell. This serves delay testing needs in which multiple stimuli and responses are to be transferred during test cycles. The 1500 standard allows a lot of freedom in the usage of Transfer in test protocols. Since there is no event in 1149.1 that is compatible with Transfer, it cannot simply be mapped onto a TAP state in the standard protocols used for TAP scan access.

In Figure 131, an example mapping of TAP states to Transfer is shown. This mapping supports the transfer of data from an off-shift-path register, that is able to capture data, to an on-shift-path register prior to shifting the data out of the WBR. An example WBR cell that captures data in a shared off-shift-path update register is shown in Figure 49 (page 139).

The first two cycles of the scan cycle shown in Figure 131 are identical to the standard JTAG scan protocol shown in Figure 130. The Transfer event must occur between the capture and the shift event. The TransferDR signal is made active in the JTAG Pause DR (PDR) state. This allows the Transfer event to be performed at the rising edge of TCK, as shown in cycle 4. WBR cells that do not support the Transfer event hold their value. In cycle 3, all WBR cells hold their value because no WSC events are triggered in this cycle. The shift state (SDR) is reached via the E2D state in cycle 5. The cycles at end of the protocol are identical to a standard JTAG scan protocol.

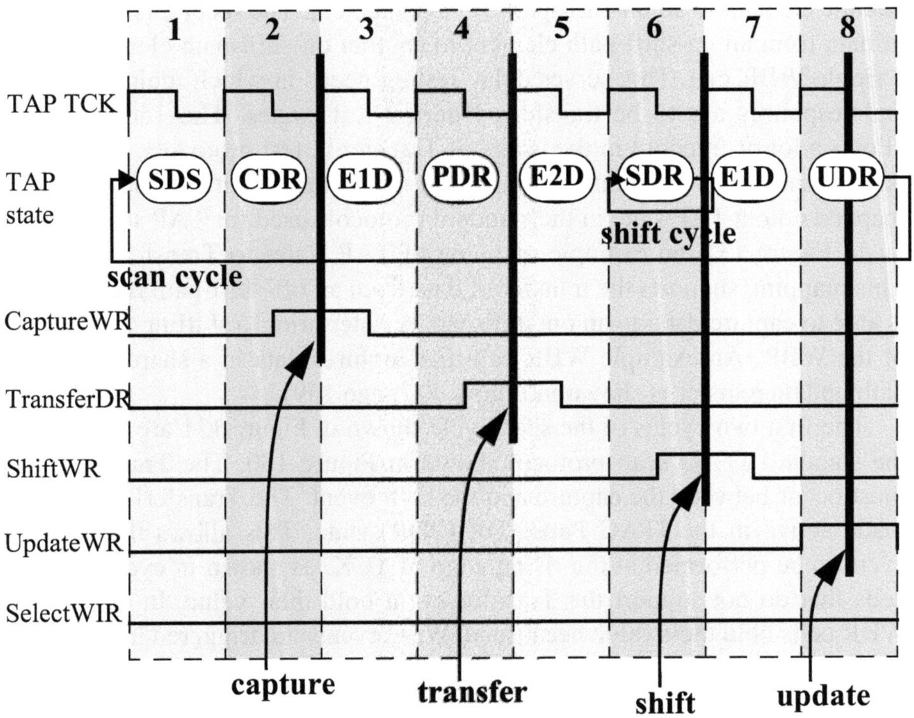

Figure 131 Example WBR cell Transfer event when controlled as JTAG data register

The example mapping shown in Figure 131 follows the mapping for the capture, shift and update events as defined by the 1149.1 standard. As a result, adding Transfer to the standard JTAG scan protocol is a limited effort. Some user-defined usage of Transfer might require a TAP interface that deviates strongly from 1149.1 and comes with a very specific protocol. In this case direct access to the WSP provides a more flexible interface without any protocol timing restrictions between desired WSC events.

11.3 Scheduling considerations

The time that is spent to test a complete SOC is an important factor in the recurring costs of testing this SOC. The type of ATE that is required to test an SOC is a second factor that heavily contributes to these costs. In general the size of the SOC test vector set determines both the required test vector memory capacity of the ATE, as well as the final test execution time. Reducing the test vector set of an SOC directly reduces the costs of testing. Minimizing this set after TAMs have been designed, wrappers have been connected, and tests have been created and translated to SOC pins, can be done by test scheduling.

Test scheduling focusses on scheduling the various SOC and core tests in such a way that no resource conflicts occur and overall test time is minimized. Scheduling can be done at various levels, ranging from individual test pattern bits, test patterns and test protocols, to scheduling of complete tests. This section discusses test protocol scheduling for 1500 compliant cores that, by definition, come with tests described in terms of patterns, protocols, and instructions.

Figure 132 shows an example test schedule for the SOC architecture shown in Figure 124 (page 260) The SOC has two cores embedded, each having a WSP and a WPP. Both cores are fully tested by scan test only. The WPPs of both cores are daisy-chained and connected to SOC pins for scan access via pin reduction logic. This logic is controlled from a WIR that is daisy-chained with the core WSPs to form a register controlled by the JTAG TAP controller.

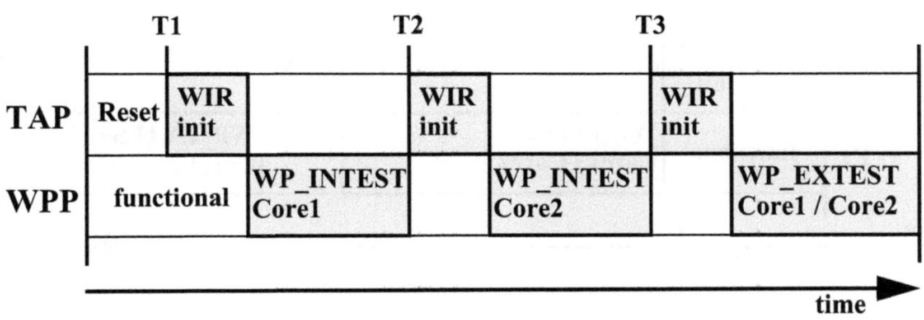

Figure 132 Example test schedule with WIR initialization and WPP protocols

To fully test the SOC, three tests must be executed. This can be done in any order. The first test is scheduled at time T1, shown in Figure 132. This test covers the internal logic of Core1 which is tested completely via the WPP. The tests starts with the initialization of the WIRs via the TAP. The WIR of Core1 is loaded with the WP_INTEST opcode. The other WIRs are loaded with instructions that enable efficient access to Core1's WPP, (e.g. WP_BYPASS). After shifting the WIR contents, the TAP controller is typically brought into Run Test Idle. In this state the WP_INTEST protocol is executed. At time T2 the internal logic test of Core2 starts as soon as the testing of Core1 has finished. The WIRs of Core2 and Core1 are loaded with opcodes for WP_INTEST and WP_BYPASS respectively. Finally at time T3 the core interconnect test is scheduled. The initialization of this test loads WP_EXTEST into both core's WIRs.

The schedule presented in Figure 132 is referred to as a 'sequential' test schedule. This type of schedule does not make use of any parallelism between tests. Power dissipation during test can be kept minimal since only one core is tested at a time. For the same reason, diagnosis of production test results is relatively easy to automate. Also, reordering tests to implement 'early abort on fail' on the production tester requires limited effort.

The daisy-chain of WPPs supports both a sequential as well as a parallel test schedule. The example test schedule shown in Figure 133 is constructed with parallel WP_INTEST protocols for Core1 and Core2. Compared to the pure sequential schedule shown in Figure 132, this schedule uses only two WIR initialization operations which results in more efficiency from a test time perspective. Executing Core1 and Core2 scan tests in parallel does not result in a significant test time decrease when compared to the sequential schedule. Testing cores concurrently typically increases power dissipation when compared to testing the cores one at a time.

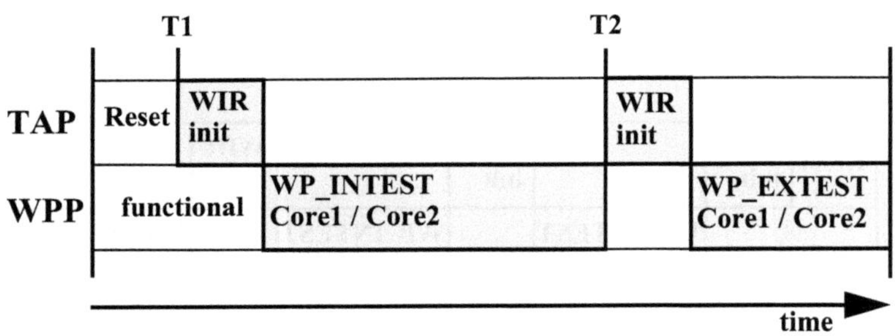

Figure 133 Example test schedule with parallel WPP protocols

One can observe in Figure 133 that the TAP and the WPP are not used continuously. The SOC test time can be further decreased with parallel load of the WIR instruction and execution of the scan protocol as shown in Figure 134.

As discussed in Section *8.1* (see page 187), new instructions can be loaded into the WIR shift register without disturbing the active test mode driven from the update register. This is also the case when the WIR is controlled from a TAP controller. This means the WIR initialization can overlap with the test protocol execution as long as no WIR update operation occurs. At time T2 in Figure 134 the interconnect test starts by shifting WP_EXTEST into the WIRs. The WIR update operation is scheduled directly after the last cycle of the combined WP_INTEST protocol execution at time T1'.

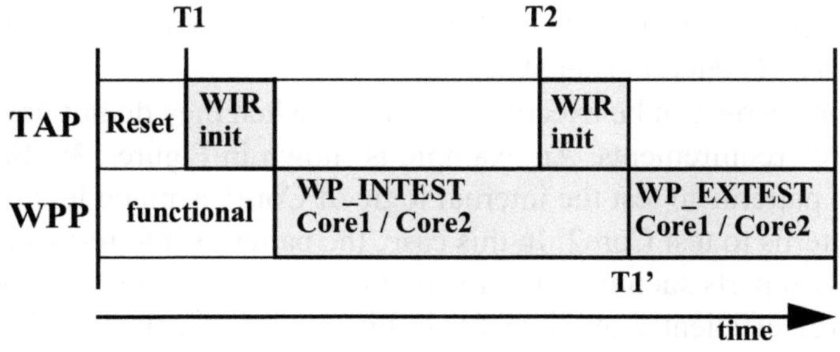

Figure 134 Example test schedule with WPP protocols and WIR initialization overlapping

Until now only test protocols using the WPP have been discussed. Suppose the two cores shown in Figure 124 (page 260) do support test of internal logic via the WPP, but no special arrangements have been made to test interconnect logic efficiently. Both cores do not support the WP_EXTEST instruction. This may not be a significant issue if the interconnect mainly consists of wiring and thus can be tested by a small amount of patterns. Figure 135 shows a possible test schedule for this case.

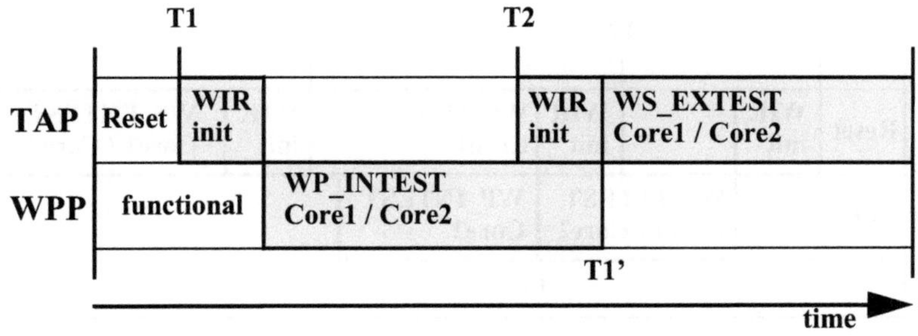

Figure 135 Example test schedule with sequence of WPP and WSC protocols

Testing the interconnect is done via the mandated WS_EXTEST instruction. Both cores load this instruction into the WIR update stage after the core internal testing has finished at time T1'. The serial interconnect test protocol, by definition, uses only the WSPs of both cores. This means that the WSPs are used for loading instructions into the WIR as well as for controlling data to and observing data from the WBR during test. Since the WSPs are fully controlled by the TAP controller care must be taken on how the SelectWIR signal is controlled. The architecture shown in Figure 124 (page 260) typically switches from WIR access to WBR access by loading a different JTAG instruction. The architecture presented in Figure 125 (page 261) typically enables WIR access during JTAG TAP instruction register states and WBR access during JTAG data register states.

Core test protocols can be executed in parallel when they do not have conflicting resource requirements. An example is shown in Figure 136. Suppose the number of patterns to test the internal logic of Core1 is much less than the number of patterns to test Core2. In this case, the pattern set to test Core2 can be divided in two parts such that the first part can be executed in parallel with the test of Core1 efficiently. At time T1, in Figure 136, the first test is scheduled. Both Core1 and Core2 execute WP_INTEST. At time T2 the WIRs are loaded with the opcodes that are activated at time T1'. The WIR of Core2 is reloaded with the WP_INTEST instruction to execute the protocol for the second part of the test patterns. The WIR of Core1 is loaded with a WS_MBIST instruction which enables the execution of memory self test via the WSP. Note that the WIR of Core2 needs reloading only if WSP control signals are shared (e.g. if the WSPs are daisy-chained as shown in Figure 124 (page 260)). At time T2' the test of Core2 has finished while the memory tests of Core1 are still running. The WIR contents for the final test can be loaded only at T3 when the memory tests are complete, since both need WSP access.

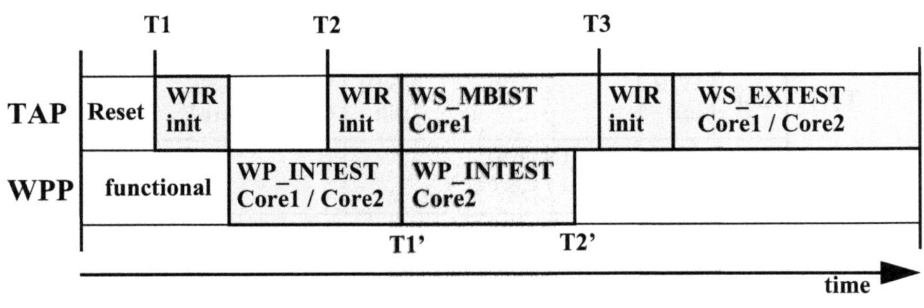

Figure 136 Example of test schedule with WPP and WSC protocols in parallel

Conclusion

This book has provided an understanding of the 1500 architecture, not only from a theoretical perspective but also from a practical perspective, by discussing implementation details relevant to the adoption of this standard. The example design, utilized throughout the book to illustrate the application of the standard, offers implementation choices that demonstrate the ample latitude built into the 1500 standard to make efficient and automated test reuse achievable, even for complex DFT strategies.

Overall, design reuse has been a significant enabler for the efficient design of complex SOC chips, in that it has allowed for reduced design cycle time. From this, it is only natural to expect that the same concept of reusability can apply to the testing, debugging and diagnosing of these chips, thereby reducing the risk of making DFT, test, debug or diagnosis the bottleneck in achieving aggressive Time-To-Market goals. The 1500 standard does make this possible.

Index

A

ActiveState keyword (CTL) 43
AssumeInitialState keyword (CTL) 43
AUXCKn 30

B

bubble diagrams 147
BYPASS 79

C

capture operation (WBR) 137
CaptureWR 30
CDRs 7
Cell Functional Input 135
Cell Functional Output 135
Cell Test Input 136
Cell Test Output 136
CFI 135
CFO 135
Compliancy (CTL keyword) 221
Core Data Registers 7
CTI 136
CTLMode block (CTL) 39
CTO 136

D

DataType block (CTL) 42
dedicated WBR cell 140
DomainReferences block (CTL) 99
DriveRequirements block (CTL) 45
dynamic test control 187

E

ElectricalProperty keyword (CTL) 40
Environment block (CTL) 39
events (WBR) 151
External block (CTL) 171
EXTEST 79

H

hierarchical cores 217
hybrid instruction 78

I

Instruction naming convention 78
Intellectual Property 11
Internal block (CTL) 40
INTEST 78
Inward Facing mode 152

M

Macro block (CTL) 103
MacroDefs block (CTL) 95
mandatory instructions 79
mission mode 31

N

naming convention (WBR cell) 149
Non-hazardous mode 153
Normal mode 152
normal mode 31, 80, 135

O

observe-only cell 166
opcode 77
Outward Facing mode 152

P

parallel instruction 78
Pattern block (CTL) 94, 101
PatternBurst block (CTL) 101
PatternExec block (CTL) 102
PatternInformation block (CTL) 100
plug-and-play 7
Procedure block (CTL) 103

R

Registered port 140

S

safe state 85
ScanDataType keyword (CTL) 44
ScanStructures block (CTL) 173
SelectWIR 30
serial instruction 78
shared storage element 138
shift operation (WBR) 136
ShiftWR 30
SignalGroups block (CTL) 27
Signals block (CTL) 27
silent shift path 84, 86, 162
static test control 187
StrobeRequirements block (CTL) 47

T

test data 11
test mode 28
test protocol 11
Transfer operation (WBR) 139
TransferDR 30

U

Unregistered port 141
unwrapped compliance 10
update operation (WBR) 138
UpdateWR 30

W

WDRs 6
WP_EXTEST 84
WP_PRELOAD instruction 87
WPI 38
WPO 38
WPP 29
wrapped compliance 10
Wrapper Boundary Register 8
Wrapper BYpass Register 9
Wrapper Data Registers 6
wrapper disabled 191
wrapper disabled state 134, 188
wrapper enabled 191
Wrapper Instruction Register 8
wrapper interface port 7
Wrapper keyword (CTL) 41
Wrapper Parallel Input 38
Wrapper Parallel Output 38
Wrapper Parallel Port 29
Wrapper Serial Port 29
WRCK 30
WRSTN 30
WS_BYPASS instruction 80
WS_CLAMP instruction 87
WS_EXTEST instruction 80
WS_INTEST_RING instruction 88
WS_INTEST_SCAN instruction 88
WS_PRELOAD instruction 86
WS_SAFE instruction 85
WSI 30
WSO 30
WSP 29
Wx_EXTEST instruction 85
Wx_INTEST instruction 82

MIX
Papier aus verantwortungsvollen Quellen
Paper from responsible sources
FSC® C105338

If you have any concerns about our products,
you can contact us on
ProductSafety@springernature.com

In case Publisher is established outside the EU,
the EU authorized representative is:
Springer Nature Customer Service Center GmbH
Europaplatz 3, 69115 Heidelberg, Germany

Printed by Libri Plureos GmbH
in Hamburg, Germany